Communication Satellite Systems: An Overview of the Technology

Edited by

R. G. GOULD
Y. F. LUM
Satellite Systems Panel
of the
IEEE Aerospace and Electronic Systems Society

Published for the IEEE Aerospace and Electronic
Systems Society by the IEEE PRESS

The Institute of Electrical and Electronics Engineers, Inc., New York

International Standard Book Number: 0-87942-065-0
Library of Congress Catalog Card Number: 75-39327

Printed in the United States of America

ACKNOWLEDGMENT

The Satellite Systems Panel of the IEEE Aerospace and Electronics Systems Society is indebted to the following authors for their contributions to this work:

Jai P. Singh, Indian Space Research Organization, Bangalore, India

James M. Janky and James G. Potter, Federation of Rocky Mountain States, Denver, Colorado, USA

W. M. Evans and N. G. Davies, Communications Research Centre, Ottawa, Canada

W. H. Hawersaat, NASA Lewis Research Center, USA

Martin P. Brown, Jr., Communications Satellite Corporation, Washington, D.C., USA

Robert K. Kwan, Telesat, Ottawa, Canada

Richard G. Gould, Satellite Systems Engineering, Inc., Washington, D. C., USA

Marcel R. E. Bichara, Vibro-Meter, AG Fribourg, Switzerland

Leo M. Keane and David W. Lipke, COMSAT General Corporation, Washington, D. C., USA

H. L. Werstiuk, Communications Research Centre, Ottawa, Canada

Mike A. Proctor, Department of National Defense, Ottawa, Canada

P. Rossiter, Telesat, Ottawa, Canada

Nelson McAvoy, Goddard Space Flight Center, Greenbelt, Maryland, USA

Donald Jansky, Office of Telecommunications Policy, Washington, D. C., USA

J. D. Palmer, Department of Communications, Ottawa, Canada

E. R. Walthall, RCA Astro-Electronics Division, Princeton, New Jersey, USA

Y. F. Lum, Bell-Northern Research, Ottawa, Canada

J. H. W. Unger, Bell Telephone Laboratories, Holmdel, New Jersey, USA

Joseph Deskevich, Operations Research, Inc., Silver Spring, Maryland, USA

Ralph E. Taylor, NASA Goddard Space Flight Center, Greenbelt, Maryland, USA

The authorship of each section is identified by the name of the author at the beginning of the section.

PREFACE

The Satellite Systems Panel of the IEEE Aerospace and Electronics Systems Society publsihed "A Review of Satellite Systems Technology" in 1972. The field of communications satellites has not remained static—in the years since that publication, many satellites have been launched and many more are being planned and designed. The Panel therefore decided that an updated version of the *Review* should be published; this volume and its companion are the result. They incorporate many of the suggestions received upon publication of the previous work. A major change is the title; it is now "Communication Satellite Systems: An Overview of the Technology". Another change is recognition of section or sub-section authors.

This publication should provide a reference on communication satellite systems, easily accessible and understood by scientists and engineers who may not themselves be satellite system specialists. Comments are welcome and should be addressed to the Editors.

Responsibility for the contents rests solely upon the individual authors and not upon the IEEE, the AES Society or its members; furthermore, the contents do not represent in any way the official views or policies of any of the authors' past or present parent organizations.

The companion volume *Literature Survey of Communication Satellite Systems and Technology* by J. W. H. Unger contains the material originally destined for the Bibliography portion of this volume. The large amount of material prepared by Mr. Unger necessitated the division of this publication into two volumes.

The Editors express their sincere thanks to the Editorial Committee and to the authors and their past or present parent organizations; to Dave Dobson, Administrative Editor of the IEEE/AES Society and to all others who have made contributions to this publication.

EDITORIAL COMMITTEE

Jai P. Singh, Chairman, Communication and Broadcast Satellite Systems Committee

Edward R. Walthall, Chairman, Satellite and Earth Station Technology Committee

Joseph Deskevich, Chairman, Navigation, Observation and Data Relay Satellite Systems Committee

Donald Jansky, Chairman, Spectrum and Orbit Utilization Committee

J. H. W. Unger, Chairman, Publication Committee

EDITORS:

Dr. Y. F. Lum, Bell Northern Research, P.O. Box 3511, Postal Station "C", Ottawa, Canada, K1Y 4HF

R. G. Gould, Satellite Systems Engineering, Inc., 7315 Wisconsin Avenue, Washington, D. C. 20014

INTRODUCTION

The state of the art in communication satellites is described in the voluminous technical literature of the various Groups and Societies of the IEEE and in the publications of the professional societies of many countries. This work is intended as an overview of the field. It describes past, present and future communications satellites, and discusses the problems faced by system designers. The companion volume is a guide to the literature. The material contained herein, together with the extensive references, should give the reader access to any desired level of communication satellite systems technology.

In this volume there are five parts and an Addendum. Part I discusses the communications satellite systems that have been used or are planned for the near future. A wide range of programs is covered, ranging from the experimental ATS-6, CTS, OTS, SYMPHONIE and Japan's experimental communication satellite projects, to the commercially operational systems of INTELSAT, TELESAT Canada, the U. S. domestic satellite networks, MARISAT and the USSR's MOLNIYA. Also covered are mobile and thin-route systems designed to supply a variety of telecommunications services to remote areas.

Satellite systems, as with any communications medium, have to be efficient and economic to be successful. Techniques for optimizing and expanding communication satellite system capabilities are therefore of major interest and are discussed in Part 2.

Frequency spectrum and orbital space are scarce natural resources. Their management and utilization are reviewed in Part 3.

Part 4 deals with earth-station technology, an area in which rapid changes are foreseen. Early earth stations had a wide range of capability, much flexibility and a great deal of equipment redundancy, and were therefore expensive. Later designs were tailored to specific system requirements. Future earth stations are likely to be increasingly specialized and to incorporate improved space-segment technology (such as high-power transponders), with a resulting improvement in cost effectiveness.

For satellite systems to be useful, they must be integrated with the existing world-wide terrestrial communications networks. Highlights of the interface problems are discussed in Part 5. Additional information for further reference on observation, scientific, tracking and navigation satellites is contained in an Addendum.

Since the survey will not serve as a reference for scientists and engineers (especially those who are non-specialists in satellite systems) without a comprehensive bibliography this is included in the companion volume, *Literature Survey of Communication Satellite Systems and Technology*, by J. H. W. Unger.

NOTES ON USAGE

Throughout this document, certain terms or symbols are frequently used. Their definition and, in some cases, their variants are discussed in these notes.

Gain

In the frequency range where a dipole, or an array of dipoles is a useful antenna, gain (G) is often given with reference to the gain of a dipole and expressed in decibels, dB. (The gain of a dipole is 2.15 dB above that of an isotropic radiator.)

At higher frequencies, gain is most often given with respect to that of an isotropic radiator and expressed in dB_i. Often this term is abbreviated to dB, thus leading to the possibility of confusion. Throughout this Volume gain is given with respect to that of an isotropic radiator.

Noise temperature

Noise temperature is expressed in Kelvins (K) or degrees above absolute zero ($-273.15°C$). Kelvins, and therefore the symbol K, include the concept of degrees; thus "$°K$" is redundant.

Figure of Merit

Figure of merit of a receiving system, G/T, is given in dB/K, or sometimes dBK.

Geostationary

Geostationary means having an orbital period equal to that of the earth, that is, nearly 24 hours; hence, not moving with respect to the earth. The term should not be confused with "geosynchronous," or "synchronous," which implies merely that the orbital period is some integral multiple of the earth's period as in the case of a synchronous satellite having a period of 8 hours.

Frequencies—Lettered Frequency Bands

Lettered bands are obsolete means of denoting frequency. The frequencies shown in parentheses should be used.

L - Band	(.390 GHz to 1.556 GHz)
S - Band	(1.55 GHz to 5.2 GHz)
C - Band	(3.9 GHz to 6.2 GHz)
X - Band	(5.2 GHz to 10.9 GHz)
K - Band	(10.9 GHz to 36.0 GHz)

EIRP

EIRP stands for effective isotropically radiated power expressed in decibels above one Watt (dBW). This quantity is the result of adding the radiated power in dBW to the gain of the antenna in dB¡ to which this power is delivered.

ERP

ERP stands for effective radiated power. This quantity is the result of adding the radiated power expressed in dBW to the gain of the antenna to which this power is delivered, expressed in decibels (dB) above the gain of a half-wave dipole. A possibility for confusion exists if a radiated power has been calculated using the gain of the antenna with respect to the gain of an isotropic radiator and then referred to as "ERP", leading the reader to believe that gain was referred to gain of a dipole.

Throughout this report, all radiated powers should be taken as EIRP.

Contents

Figures

Part 1: **Communication Satellite Systems**

Part 2: **Techniques for Expanding Communication Satellite System Capabilities**

Part 3: **Frequency and Orbit Coordination and Utilization**

Part 4: **Earth Station Technology**

Addendum: **Observation Experimentation, Tracking and Navigation Satellite Systems**

Part 1

Communication Satellite Systems

The ATS-6 Satellite Instructional Television Experimental (SITE) In India

JAI P. SINGH
Indian Space Research Organization
Bangalore, India

APPLICATIONS TECHNOLOGY SATELLITE-6 (ATS-6) PROJECT

Of the experimental satellite programs to date, perhaps the most encompassing has been the National Aeronautics and Space Administration's (NASA's) Applications Technology Satellite (ATS) Program. The underlying design philosophy of the ATS program has been the development of multiple mission satellites having a large and adaptable volume for mounting the various experiment payloads. Included in ATS missions 1 through 5 are Super High Frequency (SHF) voice and television experiments at 6/4 GHz, Very High Frequency (VHF) communications experiments with mobile stations, 15.3-GHz propagation measurements and an experiment in the band 1.55—1.65 GHz which provided information for use in aeronautical and maritime applications. (1—8)

ATS-6, last of the ATS series of satellites, was successfully launched from the Eastern Test Range (ETR) on May 30, 1974. The prominent features of the ATS-6 spacecraft weighing about 1360 kg (3000 pounds) are: a 10-m (33-foot) diameter, deployable, parabolic reflector with multiple offset feeds, and fine pointing ($0.1°$), slewing ($17.5°$ in 30 minutes) and tracking capability. The spacecraft communications system includes a composite feed assembly and an integrated transponder subsystem, and thus, in association with the 10-m (33-foot) diameter antenna, provides the capability for a multiplicity of communications experiments at radio frequencies ranging from VHF to millimeter waves (9). Included in the experiments with ATS-6 are experiments of 6 GHz and 4 GHz radio-frequency interference between satellite and terrestrial systems, 860 MHz Television Relay Using Small Terminals (TRUST) and Satellite Instructional Television Experiment (SITE), 2.5 GHz Educational Television experiment, Position Location

and Aircraft Communication Experiment (PLACE), 18/13 GHz Millimeter Wave (MMW) propagation measurements, and demonstration of technology necessary for an operational tracking and data relay satellite system (10).

SATELLITE INSTRUCTIONAL TELEVISION EXPERIMENT (SITE)

The SITE experiment is a joint experiment between the Department of Space (Government of India) and NASA agreed upon formally on September 18, 1969 (11). It will be conducted from July 1975 to July 1976 in India using the 860-MHz transponder on ATS-6. Its basic purpose is to demonstrate the potential value of satellite broadcast television (TV) in the instruction of village habitants and communicating educational messages to remote areas, to gain experience in the development, testing and management of a satellite-based instructional television system (particularly in rural areas) and to determine optimal system parameters (12).

After one year of service at $94°$ W longitude for experiments in the U.S., the ATS-6 satellite will be moved in early June 1975 to $34°$ E for the SITE experiments. In the SITE experiment, a frequency-modulated TV carrier (having two audio subcarriers multiplexed with the modulating TV base band) at 6 GHz will be transmitted to the ATS-6 Earth Coverage Horn (ECH) antenna from one of the two earth stations at Ahmedabad and Delhi. At the spacecraft, the received signal will be processed and retransmitted at both 4 GHz and 860 MHz. The 4-GHz downlink will be used to monitor system performance only on a limited time basis. At 860 MHz, the beam-center spacecraft effective isotropically radiated power (EIRP) is to be greater than 51.3 dBW, permitting the use of small and low-cost receive terminals.

3

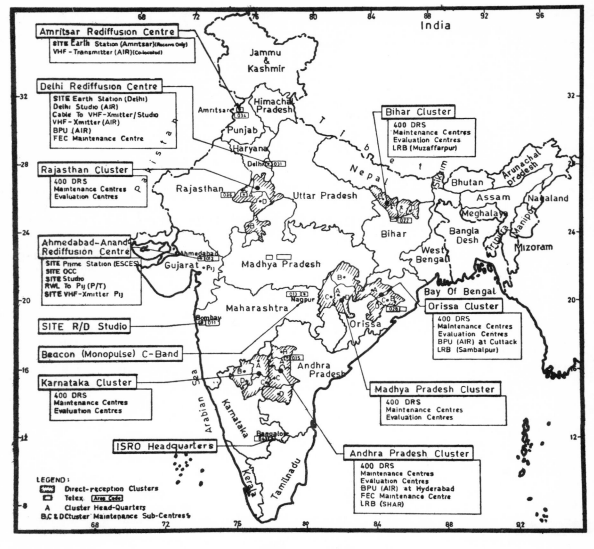

Fig. 1. SITE Centers of Activity (13)

The ground segment for SITE, in addition to receive and transmit earth stations at Ahemedabad and Delhi, is to consist of 2400 Direct Community Receivers (DCRs) located in six of the Indian states in clusters of about 400 each, two Limited Rebroadcast Transmitters (LRBs), and a receive only earth station at Amritsar co-located with the local TV broadcast transmitter. In addition, it is planned to have a 1 kW VHF TV transmitter at Nadiad in Gujarat interconnected with the Experimental Satellite Communication Earth Station (ESCES) at Ahmedabad, the prime earth station for SITE, by a terrestrial microwave link. Figure 1 gives the centers of activity of SITE in India. It also shows the composition of the ground segment of SITE now being established.

DCRs in SITE are to have a 3-m (10-foot) diameter chicken-mesh antenna and a system G/T of −6 dB/K allowing for 1 dB degradation in G/T

with time. The LRBs are to use 5-m (16-foot) diameter antennas and a lower-noise RF front end, as compared to that of the DCRs, to receive the 860-MHz satellite signal and then rebroadcast the signal in the VHF band to conventional receivers within 12−15 km range.

References

1. TRW Space Log, vol. 9, no. 2, Summer-Fall 1969.
2. "Super High Frequency (SHF) Communications System Performance on ATS," vol. 1, System Summary, NASA-TM-X-65304, National Aeronautics and Space Administration, Goddard Space Flight Center, Greenbelt, Maryland, August 1970.
3. "ATS-1 VHF Communications Experiment Final Report, May 1967−December 1969," FAA-RD-70-12, National Aviation Facilities Experimental Center, Atlantic City, New Jersey.

4. L.J. Ippolito, "Millimeter Wave Propagation Measurements from the Applications Technology Satellite (ATS-5)," *IEEE Transactions on Antennas and Propagation,* vol. AP-18, no. 4, July 1970.

5. F.J. Kissel, "L-Band Performance Characteristics of the ATS-5 Spacecraft," *IEEE International Conference on Communications,* San Francisco, June 1970.

6. O.J. Hanas, and R.M. Wae-jen, "Navigation and Communication Experiment, at L-Band onboard S.S. Manhattan using ATS-5 Satellite," *IEEE EASCON Proceedings*, Washington, D.C., October, 1970, pp. 132–155.

7. L.J. Ippolito, "Effects of Precipitation on 15.3 and 31.65 GHz Earth-Space Transmissions with the ATS-V Satellite," *IEEE Proceedings*, vol. 59, no. 2, pp. 189–205, February 1971.

8. "Applications Technology Satellites—A Continuing Bibliography with Indexes," National Aeronautics and Space Administration, NASA-X-560-72-87, Goddard Space Flight Center, Greenbelt, Maryland, March 1972.

9. A.B. Sabelhaus, "Applications Technology Satellite F and G Communication Sub-System," *IEEE Proceedings*, vol. 59, no. 2, February 1971.

10. "The ATS-F and-G Data Book," National Aeronautics and Space Administration, Goddard Space Flight Center, Greenbelt, Maryland, September 1972.

11. J.P. Singh and D.T. Jamison, "The Satellite Instructional Television Experiment in India: A Case History," Center for Development Technology, Washington University, Saint Louis, Missouri, July 1973.

12. "Memorandum of Understanding between the Department of Atomic Energy of the Government of India and the United States National Aeronautics and Space Administration," National Aeronautics and Space Administration Headquarters, Washington, D.C., September 18, 1969.

13. "Annual Report: 1974-76," Department of Space, Government of India, Bangalore, India, 1975.

The ATS-6 Health Education Telecommunication (HET) Experiment

JAMES M. JANKY
JAMES G. POTTER
Federation of Rocky Mountain States
Denver, Colorado, U.S.A.

ACKNOWLEDGEMENT

This paper was written with the assistance of Messrs. Al Whalen and Eldon Volkmer of the Goddard Space Flight Center and the entire staff of the Broadcast and Engineering Component of the Federation of Rocky Mountain States. Special thanks are due to Gordon Law and Dail Ogden, the Directors of the Satellite Technology Demonstration and the Broadcast and Engineering Component, respectively. Also, we wish to express our appreciation to Ms. Jean Braunstein, who compiled all of the materials used in this paper.

INTRODUCTION

On May 30, 1974, the National Aeronautics and Space Administration (NASA) launched the ATS-F satellite into a geostationary orbit from Cape Kennedy, Florida, at which time it was redesignated as ATS-6. The satellite is situated above the equator at a longitude of 94° West providing communications services to experimental groups in North America. In summer of 1975 it will be moved to India for the Indian experiment with broadcast television to a large number of villages.

Using the ATS-6 as the prime satellite, along with 2 previously launched satellites, the ATS-3 and the ATS-1, the Department of Health, Education and Welfare, NASA, and the Corporation for Public Broadcasting are jointly sponsoring one of the principal experiments called the Health Education Telecommunications (HET) experiment. The purpose of the HET is threefold: (1) Demonstrate a satellite TV distribution system that could be implemented commercially at a cost that will ensure its usefulness to such public services as health and education; (2) Explore technical and organizational mechanisms for dealing simultaneously with the need for high quality audio-visual materials at low per capita cost and the desire to individualize services to meet specific local needs; (3) Develop several technology-based system models in service areas where public commitment is evident but no developed institutional response exists (1).

Six independently managed experiments constitute the HET network which covers 23 states in the Rocky Mountain, Appalachian, and Alaskan (northwest) regions. Each experiment is unique, having variations in program, audience, and equipment configuration.

The Indian Health Service (IHS), a branch of the Health Services Administration, is conducting experiments in the delivery of health care and continuing medical education, principally at sites in Alaska. The IHS Office of Systems Development and Analysis at IHS headquarters in Rockville, Maryland, and its Alaska Area Native Health Service office in Anchorage are participating.

The Washington, Alaska, Montana, Idaho Program (WAMI), with headquarters at the University of Washington, Seattle, is engaged in the improvement of medical education in those states and has coordinating offices at the state universities of the latter three states as well. WAMI will conduct experiments in medical curriculum, administration, computer-aided evaluation and community health in Washington and Alaska.

The Alaskan education experiment, which is under the direction of the Office of Telecommunications within the Office of the Governor of Alaska, will provide culturally relevant educational and health training and information to various groups including small children, students, paraprofessionals and other adults. Additionally, real-time public broadcast programming (Public Broadcast System and National Public Radio) will be delivered to Alaskan populations.

A fourth experimenter in the HET network, the Veterans Administration, is performing a series of biomedical communications experiments in which consultation and continuing education are provided to VA hospitals in the Appalachian region from major medical centers.

The Appalachian Educational Satellite Project (AESP), under the direction of the Appalachian Regional Commission, is an experiment in the delivery of training programs in elementary reading and career education to Appalachian educators (teachers, counselors, administrators). A Resource Coordinating Center (RCC) has been established at the University of Kentucky in Lexington, Kentucky. The RCC develops the courses, seminars and information systems, including a computer retrieval system, which are delivered through Regional Education Service Agencies (RESA's) to the participating educators at variously located terminals.

The Satellite Technology Demonstration (STD), managed by the Federation of Rocky Mountain States (FRMS), is delivering career education courses to rurally located junior high school students and programs for general community audiences which cover a variety of topics. Larger audiences in that Rocky Mountain region are able to view the various programs through special arrangements and installation of receiving equipment at public broadcast stations. Additionally, a Materials Distribution Service is offered; i.e., request programs selected from a special catalog are transmitted via satellite at designated times so that they can be recorded on videocassette. The STD is a prototype project designed to provide information on the feasibility of telecommunication satellite systems for the delivery of social services to selected populations. Therefore, a major task of the Demonstration is the assembling of a data base to assist decision makers on questions of resource allocation and other policy matters regarding satellite technology. Investigations will determine: (1) the effectiveness of the total demonstration system in delivering programs to rural audiences; (2) the effectiveness of field support and public information in attracting and holding audiences; (3) the impact of programming and related support services on individuals in the viewing audience; (4) the impact of the delivery system and its products on various participating agencies; (5) the major factors associated with development, implementation, and evaluation of the demonstration; and (6) the costs associated with these factors.

The Broadcast and Engineering Component of the Satellite Technology Demonstration has been designated to serve as operations manager of the HET ground network, which includes all six experiments. Working closely with NASA's Goddard Space Flight Center, Broadcast and Engineering has engineered the network of ground stations that is being used. The nature of the ground communications system, including its relationship to the satellite configuration, is also discussed.

FUNCTIONAL DESCRIPTION OF HET NETWORK

The ATS-6 is equipped with a spot-beam antenna and a global-beam antenna. Special switching capabilities of the spacecraft permit signals from 2.25 or 6 GHz uplinks to be received, processed, and transmitted to earth at either 2.6 or 4 GHz. The satellite can receive video transmissions at 6 GHz or 2.25 GHz and can transmit simultaneously a video signal at 4 GHz and at two frequencies in the neighborhood of 2.6 GHz. This flexibility affords the opportunity for the Network Coordination Center (NCC) in Denver to monitor all transmissions and to coordinate the unique distribution needs of each experiment while meeting requirements established by the Federal Communications Commission.

The bulk of the HET network experiments are designed to make use of the comparatively high effective isotropically radiated power (EIRP) of the ATS-6. With 9-m (30-foot) diameter unfurlable parabola and 15 W (minimum at end of life) of RF power at 2.6 GHz, the EIRP is 48 dBW at beam edge. This power level permits reception of high-quality color video signals at terminals with modest system sensitivities. The range of figure of merit (G/T) which will provide TASO 1 or better picture quality for the terminals in the HET experiment is 3.5 to 7.1 dB/K.

The HET ground segment consists of the ATS Operations Control Center (ATSOCC) at the Goddard Space Flight Center, NASA earth stations in Rosman, North Carolina, and Mojave, California, an STD earth station in Morrison, Colorado, the HET Network Coordination Center (NCC) operated by the STD in Denver, Colorado, regional control centers in Lexington, Kentucky, and Fairbanks, Alaska, and 118 terminals configured in one of 3 modes.

Terminals designed to receive and demodulate a single video signal and four associated audio channels, each transmitted as subcarriers above the video spectrum, are designated "receive-only terminals" (ROT's). The composite spectrum of the audio subcarriers and the video signal is transmitted via wideband FM on either of two RF channels whose carriers are centered at 2566.7 MHz and 2667.5 MHz. Thus, the ROT is essentially a converter which transforms the audio and video signals from FM into a standard baseband format for display on a color monitor, with a maximum of four simultaneous audio program channels available. The ROT is being used at all 118 remote sites in the HET network.

An intensive terminal consists of the basic ROT and a VHF transmitter/receiver system, which uses either the ATS-1 or the ATS-3 satellite. The VHF transmitter/receiver system provides "talk-back" audio/data link capabilities between designated sites and always includes the NCC. Three of the HET experimenters, the VA, AESP, and the STD, have a unique arrangement for the use of the VHF system. In each case the arrangement was developed on the basis of suitability to enhance the program objectives of the particular experiment.

All remote terminals in the VA experiment are ROT's located at hospitals. However, a single mobile van, using VHF interactive voice features, travels between hospitals, and on designated days a particular hospital uses two-way voice for consultation with medical experts at the KMGH studio in Denver, Colorado. The VA scheme is augmented by land-line links which provide terminals with additional two-way verbal communication and slow-scan visual transmission capabilities.

The AESP consists of five geographical clusters. Each cluster has an intensive terminal which is a Regional Education Service Agency (RESA), and two ROT's at school locations. The ROT's have teletype systems, which are conventionally routed through land-line connections to the RESA intensive terminal. The RESA intensive terminal also has a teletype system, but it is channeled via the VHF system to the ATS-3. In this way a communications link has been established between the RESA's and the Resource Coordination Center (RCC) in Lexington, Kentucky, which also is an intensive terminal.

The STD has four ROT's at school locations in each of eight states. Twelve ROT's are being installed at public broadcast stations in that region. These 44 terminals have no additional capabilities. Three STD terminals in each of eight states are intensive and have VHF systems which are routed through the ATS-3 to the Network Coordination Center in Denver. These 24 terminals have a voice and limited data capability. In the automatic data mode, digital pads will be the primary medium for transferring information between sites and the NCC.

The third HET ground terminal configuration is a comprehensive terminal which, as the designation suggests, includes ROT and intensive terminal capabilities along with additional originating video features. The terminals can transmit and receive from the ATS-6 one composite signal consisting of a television channel of 4.2-MHz bandwidth and four voiceband channels of 10-kHz bandwidth each. These terminals transmit at 2247.5 MHz and receive at 2566.7 MHz.

The sites in the Alaskan region are either intensive terminals or comprehensive terminals and are structured for flexibility of use between experiments. Therefore, the terminals may be discussed as a unit, but it must be noted that there are variations in the distribution possibilities, which, as in the case of the other experiments, are determined on the basis of the individual program needs of each experiment.

All comprehensive terminals are located in the Alaskan region and are used as part of one or more of the three experiments in that region; i.e., IHS, WAMI, or the Alaskan Education Experiment. Use of the four voiceband channels varies not only between experiments, but also between particular

programs within each experiment. A segment of one of the WAMI experiments, for example, utilizes teletype via a spare audio channel (one in each direction) for a computer-aided evaluation experiment in which medical students at the University of Alaska (Fairbanks) interact with instructional computer programs at the University of Washington in Seattle. In another instance, the Alaskan Education Experiment plans at times to deliver National Public Radio programs to Alaskan populations via one of the spare audio channels, while simultaneously delivering educational training programs to specific sites using video and one audio channel.

The intensive terminals in the three experiments primarily use the two-way voice mode (rather than the data mode) in the VHF system. The VHF system is used to coordinate all Alaskan experiments through the Alaskan Operations Center (AOC) in Fairbanks, which is the regional control center for that area.

The HET Network Coordination Center in Denver is the nerve center for all HET experimental activities. Special capabilities built into the center and made possible by the Morrison earth station allow the NCC Supervisor to coordinate all contacts between the HET ground network and NASA's ATS Operations Control Center (ATSOCC) which has control over the satellites used in the HET network. All ATS-6 video transmissions in the HET network, regardless of the originating location, can be monitored by NCC through the Morrison earth station via the 4-GHz transponder. The NCC Supervisor can exercise control over all ATS-3 and ATS-1 transmissions by having the capability to enable or disable all VHF transmissions. This is accomplished by transmitting digital commands to special logic circuits incorporated in the VHF terminal equipment. The digital control equipment is called a "Digital Coordinator." Additionally, station identification signals are transmitted to NCC via the Digital Coordinator every time a message is transmitted. All operative two-way sites are displayed at NCC on an illuminated map board.

In a 15-minute checkout period which precedes each broadcast slot, four "truth" sites in the region about to receive the programming are polled automatically by the NCC to determine their received signal strength. These sites in all cases are located along the periphery of the coverage area. Pointing errors of the 10-m (33-foot) antenna on ATS-6 result in much higher or much lower signal-strength readings at the edge of the nominal coverage area. The readings from the truth sites are processed at the NCC, and ATSOCC is provided a quantitative estimate of the required correction in spacecraft pointing if, in fact, there has been an error.

Video programming to be distributed to STD audiences in the Rocky Mountain region is transmitted from the NCC through a 12-GHz microwave relay link to the Morrison earth station, which then accesses the ATS-6 for further transmission to the terminals.

Video for the VA experiment is routed from KMGH in Denver via land lines to NCC, through the 12-GHz microwave link to the Morrison uplink facility.

Video programming for the other HET experiments is distributed from other locations and involves either direct transmission between sites, as in the case of Alaskan comprehensive terminal activities, or distribution from regional control centers (Fairbanks, Alaska or Lexington, Kentucky).

In the case of the RCC at Lexington, Kentucky, video is transmitted via microwave to NASA's earth station at Rosman, North Carolina, which then accesses the satellite.

Fairbanks, Alaska, has two terminals 6.4 km (4 miles) apart. One is a Comprehensive terminal located at the University of Alaska that can access the ATS-6 directly. This terminal is the regional control center for that region and is designated the Alaskan Operations Control Center (AOC). The other location is an Intensive terminal at the Fairbanks Indian Health Center. This terminal transmits video via 9-km (5½-miles) microwave link to the AOC, which in turn accesses the satellite from that point.

If failures occur, video routes are temporarily restructured. Contingency actions have been developed by each experiment whereby through mutual agreement one station acts as a backup for another. Videotaped copies of other experiments' programs are on hand for broadcasting in the event of an uplink failure.

The Morrison earth station, located 21 km (13 miles) southwest of Denver, can access the ATS-6 in any one of three 30-MHz bands centered at 5959 MHz, 6150 MHz, or 6350 MHz. With one exception, reception of HET programs via ATS-6 at the earth station will be accomplished at 4 GHz via the global beam on the spacecraft. As Morrison is located within the same eastern antenna beam coverage as the STD remote terminals, STD programs received in the east mode at 2.6 GHz from the spacecraft's spot-beam also will be received at Morrison on that frequency.

Transmissions between the earth station and the ATS-3 and ATS-1 will occur on 149.22 MHz for the uplink and 135.6 MHz for the downlink. A dedicated phone line between Morrison and the NCC relays these VHF transmissions.

DESCRIPTION OF HARDWARE

Receive-Only Terminal

ROT equipment includes a 3-m (10-foot) diameter parabolic reflector, a prime-focus feed with a low-noise preamplifier located directly at the feed, an interconnecting cable, an indoor unit which serves as a demodulator, and a color television monitor. The antenna is supported by an adjustable mount and is circularly polarized. Figure 2 is a diagram of a Receive-Only Terminal. The indoor and outdoor units of the ROT are separately displayed.

Some terminals are equipped with an indoor elevation adjustment mechanism. The mechanism consists of a small electric motor in a weatherproof container, a gearbox, a worm-gear drive, and two limit switches to limit the adjustment range to ±3° about a nominal position.

The receiver system is manufactured by Hewlett-Packard. This system is a tuned-radio-frequency, or "TRF," receiver, so named because the demodulation process occurs directly at 2.6 GHz. Approximately 120 dB gain is provided by the amplifier chain. A fast AGC loop of unique design provides the amplitude-modulation suppression necessary for realizing the full FM advantage. The 4.2-dB noise-figure amplifier chain and discriminator cover the full 2.5–2.69 GHz bandwidth allocated for educational satellites. Either of two interdigital filters may be manually connected to provide channel selection.

A deemphasis filter and video amplifier provide a studio-quality signal at the standard 1.0 V level from a 75 Ω source. The four audio subcarrier demodulator circuits employ an integrated circuit as the key component. The output level is 0 dBM from balanced 600 Ω sources.

The television set is a standard monitor with provision for either baseband inputs or RF inputs.

Intensive Terminal

In addition to ROT equipment, the intensive terminals comprise the following basic VHF items: (1) a circularly polarized helical antenna with its mount and pad; (2) a low-noise VHF preamplifier; (3) a diplexer; and (4) a VHF transmitter/receiver with a digital coordinator. A block diagram of the intensive terminal is shown in Figure 3.

The digital coordinator is a multi-functional digital transmitting/receiving unit used to coordinate communications on the VHF system. It consists of an automatic data generator, transmit control logic, receive control logic, a parallel-serial converter, and a control panel. Data is transmitted asynchronously in the ASCII format, providing compatibility with other peripherals for which moderate quantities of data must be transmitted. The data rate for the system is limited to 1200 bit/s to eliminate the need for conditioned lines and to provide a low error rate (on the order of 10^{-4}) over the VHF satellite link.

A five-word preamble accompanies each transmission from a remote site. The first word is an ASCII escape character, and the second word is the station-identification character. The third word describes the mode of operation of the station, either "call" (a burst mode in which a channel is requested from the NCC), "voice," or "data." The fourth word is the same as the third word to reduce errors. The fifth word is an end-of-text character.

The same five-word format is transmitted from Denver to all remote sites. The third word contains a command which is executed only if the

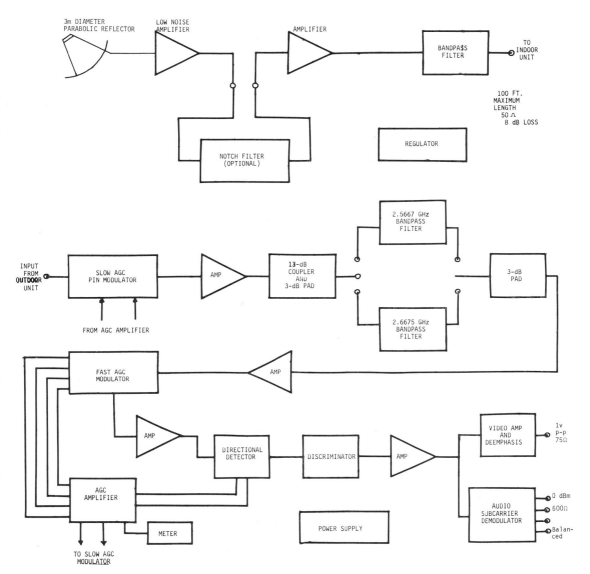

Fig. 2. The Receive-Only Terminal

proper station identification is decoded first. A typical command enables transmitter operation in the "voice" or "data" mode. Such strict discipline is necessary to prevent accidental interruptions during data transmissions from other sites, and to facilitate a quick shut-down in the event of emergency or unauthorized transmission.

Comprehensive Terminals at 2 GHz

The comprehensive terminal consists of all of the previous equipment, plus a wideband FM video transmitter, and a separate transmitter antenna. The video transmitter is capable of accepting up to four simultaneous audio channels

as subcarriers, and therefore can transmit a signal identical to the main transmissions from Denver. The transmitter consists of an indoor unit and an outdoor unit. The indoor unit accepts baseband video and audio, provides preemphasis, generates the composite video spectrum, and contains the power supply. Video is carried by cable to the outdoor unit, which contains a wide-band FM modulator and a 20-W solid-state power amplifier.

Denver Earth Station

The 4/6-GHz Denver earth station in Morrison, Colorado, is designed for simplicity and low cost.

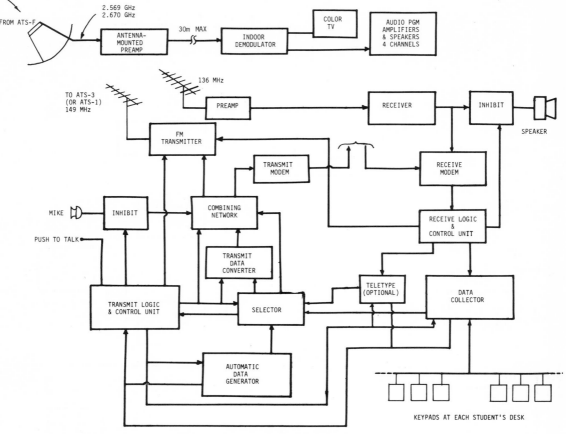

Fig. 3. Intensive Terminal Block Diagram

To obtain the required EIRP (84 dBW at 6 GHz) and G/T (28.9 dB/K at 4 GHz), a 3-kW transmitter, an uncooled parametric amplifier, and an 11-m (36-foot) prime focus parabolic antenna are used.

Changes in the pointing coordinates necessitated by the movement of ATS-6 are accomplished manually. The predicts are provided by a computer in the NCC. The terminal was supplied by Radiation Systems, Inc., of McLean, Virginia.

SUMMARY

The Health Education Telecommunications Experiments provide a range of communications capabilities to remote areas: (1) the Receive-Only Terminals deliver one-way color video plus four associated audio channels; (2) the Intensive terminals have a Receive-Only Terminal and a two-way audio/data channel with a limited digital response system; (3) sites in the Alaskan region can transmit color video for remote medical diagnostics and health and educational information purposes as well as provide all the previous capabilities.

The goal of the HET experiment is to demonstrate the feasibility of a satellite communications system, to evaluate user acceptance, and to develop cost data on the various communication system configurations.

References

1. National Aeronautics and Space Administration, *Network Operations Support Plan for ATS-6,* vol II Experiments, pg. 34-1, April 1974.

Bibliography

Alaskan ATS-F Health/Education Telecommunications Experiment, *Program Plan,* March, 1974.

Alaskan ATS-F Health Experiment Plan, University of Washington, Seattle, Washington, October 1973.

Satellite Technology Demonstration, *Experimental Test Plan,* Federation of Rocky Mountain States, Denver, Colorado, September 17, 1973.

Satellite Technology Demonstration, *Broadcast and Engineering Training Manual for Health Education Telecommunications Network Site Operators,* Federation of Rocky Mountain States, Denver, Colorado, June 7, 1974.

The Communications Technology Satellite (CTS) Program

W. M. EVANS
N. G. DAVIES
Communications Research Centre
Canada

W. H. HAWERSAAT
NASA Lewis Research Center
U.S.A.

INTRODUCTION

On April 20, 1971, an agreement between Canada and the United States was announced which established a joint Communications Technology Satellite (CTS) Program. Under this agreement Canada undertook to design and build the spacecraft while the United States undertook to provide the launch vehicle, a high-power travelling-wave-tube amplifier, its power conditioner and thermal control, spacecraft environmental test support, and launch and operational support to place the spacecraft in synchronous orbit. Both countries agreed to carry out an experimental program in communications and provide a variety of ground terminals for this purpose. NASA's Lewis Research Center and the Communications Research Centre of the Canadian Department of Communications were designated as the two project offices for the program. In May, 1972, an agreement was signed between Canada and ESRO (European Space Research Organization) wherein ESRO was to provide two 20-W travelling-wave-tube amplifiers, a parametric amplifier, and to develop an extendible solar blanket and associated solar cells.

The principal technological objectives of the program are to:

1. Conduct satellite communication systems experiments using the 12- and 14-GHz bands and low-cost transportable ground terminals;

2. Develop and flight test a super-efficient power amplifier tube having greater than 50% efficiency with a saturated power output of 200 W at 12 GHz;

3. Develop and flight test a lightweight extendible solar array with an initial power output greater than 1 kW;

4. Develop and flight test a 3-axis stabilization system to maintain accurate antenna boresight positioning on a spacecraft with flexible appendages.

The satellite will be launched in 1975 into a synchronous orbit by means of a Delta 2914 launch vehicle and a self-contained apogee boost motor. The spacecraft is intended to operate at a longitude of 116°W for a period of two years and will produce effective isotropically radiated power levels significantly greater than those provided by existing or project-approved spacecraft.

FEATURES OF MISSION DESIGN

The launch of the CTS spacecraft will follow a fairly standard Delta synchronous orbit profile. Because the Delta third stage and the spacecraft apogee motor will be fired without active attitude control, the spacecraft will be spin-stabilized from the conclusion of second-stage firing until it is located on station. In this spinning mode the deployable solar arrays will be stowed within the roughly cylindrical spacecraft. A 90-watt circumferentially mounted body solar array and two 5-ampere-hour NiCd batteries will furnish power for the spacecraft.

Following location on station, the spacecraft will be despun by low-thrust hydrazine thrusters; two sections of the body array will be jettisoned; the flexible array will be deployed; the momentum wheel will be spun up; the spacecraft will be oriented as shown in Figure 4; and the on-board 3-axis stabilization system will be activated. The spacecraft will not be capable of north-south stationkeeping and thus the initial node and inclination of the synchronous orbit will be chosen in order to maximize the number of days during the

13

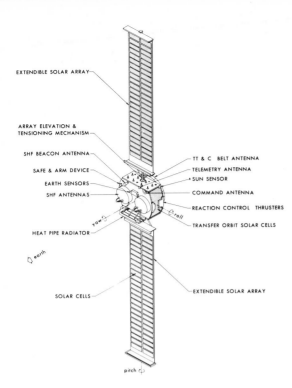

EXTENDIBLE SOLAR ARRAY

ARRAY ELEVATION & TENSIONING MECHANISM

SHF BEACON ANTENNA

SAFE & ARM DEVICE

EARTH SENSORS

SHF ANTENNAS

HEAT PIPE RADIATOR

SOLAR CELLS

TT & C BELT ANTENNA

TELEMETRY ANTENNA

SUN SENSOR

COMMAND ANTENNA

REACTION CONTROL THRUSTERS

TRANSFER ORBIT SOLAR CELLS

EXTENDIBLE SOLAR ARRAY

yaw   roll

earth

pitch

Fig. 4. Communications Technology Satellite.

*Registered trademark of SPAR Aerospace Products Ltd., Toronto, Canada.

Spacecraft Description

Figure 4 shows the spacecraft in its synchronous orbit configuration. The on-board attitude control system will maintain the extendible solar arrays perpendicular to the sun line (except for sun declination angle) and the yaw axis pointing to the center of the earth (within ±0.1°). Table 1 summarizes the major system design parameters.

The on-station attitude control subsystem is based on the "momentum bias" principle and uses a fixed momentum wheel to provide gyroscopic stiffness in the plane of the orbit. Pitch control is obtained by varying the speed of the momentum wheel while low-thrust hydrazine thrusters provide control torques about the roll and yaw axes. Pitch and roll error signals are obtained from a static infrared earth sensor and sun sensors are used to provide periodic yaw attitude information.

Each extendible solar array is stowed in a flat pack and will be extended in a concertina fashion by a stainless steel BI-STEM* boom. Each array will be 6.5-m (256-inches) long, 1.3-m (51.6-inches)

two-year lifetime that the inclination is less than 0.65°. East-west stationkeeping will be performed to maintain the spacecraft position at 116°W ±0.2°.

TABLE 1. Communications Technology Satellite Major System Parameters

Weight at liftoff	672 kg (1481 lbs)
Attitude control accuracy	±0.1° in roll and pitch ±1.1° in yaw
SHF antenna boresighting accuracy	±0.2°
Telemetry channels	373 measurements once per second
Command channels	255
Initial power from flexible array	1260 W

wide, and will contain 12,636 solar cells mounted on a thin substrate of Kapton and fiberglass. The arrays will be instrumented with strain gauges and accelerometers to provide information on flexible body dynamics and the accuracy of the three-axis attitude control system.

CTS Communications Subsystem

The communications subsystem (Figure 5) consists of two steerable antennas of 2.5° beamwidth, a 200-W TWTA, a redundant pair of 20-W TWTA's, a high-sensitivity, high-gain receiver (with a parametric amplifier as one of the redundant front ends) and an earth-coverage SHF beacon. Table 2 is a summary of the specifications for the communications subsystem. Figure 6 shows the frequency plan for the subsystem. The signals received by the two antennas in receiver bands RB1 and RB2 are combined in the input multiplexer, amplified in the low-noise receiver and frequency translated to transmit bands TB1 and TB2 respectively. In the primary mode, one of the redundant 20-W TWT's is used both as the output amplifier for TB2 and the low-power driver amplifier for the 200-W TWT in TB1. An FET amplifier is used as a driver for the 20-W TWT in band TB2.

A secondary mode is available wherein the 200-W TWT is switched off and one 20-W TWT is used as the output amplifier for both TB1 and TB2. The power split between TB1 and TB2 depends upon the relative frequencies and input power of the received signals in RB1 and RB2.

Figure 7 shows some typical coverage areas for five antenna boresight locations. In each case the outer contours represent 2.5° beamwidth coverage areas and the inner contours show the reduced coverage resulting from the stated spacecraft position and attitude errors.

Communications Experiments

The transponder is adaptable to the following types of communications:

1. TV Broadcast
   - TV broadcast to small communities in remote areas from fixed or transportable terminals

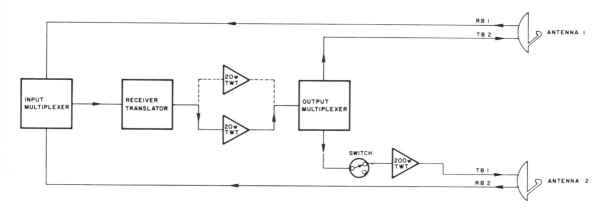

Fig. 5. CTS SHF Communications Subsystem.

TABLE 2. CTS Communications Subsystem Specifications

SHF Antennas	2.5° circular cross-section beam two-axis steerable to place electrical boresight anywhere within a 17° cone minimum transmit gain within beam: 33.6 dB minimum receive gain within beam: 33.2 dB

Transponder	2 rf channels of 85 MHz separated by 110 MHz
	transmit at 12 GHz
	receive at 14 GHz
	single heterodyne configuration
	gains available — band RB1/TB1 — 107, 112, 117, 122 dB
	                 — band RB2/TB2 — 110, 114 dB
	input noise temperature — with parametric amplifier < 1000 K
	                             — without parametric amplifier < 2000 K
	power output — primary mode 200 W (TB1) and 20 W (TB2)
	               — secondary mode 20 W split between TB1 and TB2

SHF Beacon	operating frequency 11.700 GHz
	power output — primary mode > 200 mW
	               — secondary mode > 12 mW
	antenna — boresight 20°N, 105°W
	         — right hand circular polarization
	         — > 16.5 dB gain within 17° beamwidth

- Educational TV (ETV) with a voice or data return channel for interactive programs
2. TV Remote Transmission
- TV transmission of special events in isolated areas from a transportable terminal to a central area for networks distribution or for retransmission to a remote region
3. Radio Broadcast
- Broadcast of radio program material to small ground terminals
4. Two-Way Voice
- Telephony service, including voice, facsimile and data, to and between small transportable ground terminals
5. Digital Communications
- Digital data transmission and exchange
- Investigation of high-speed data transmission by satellite

- Investigation of time division multiple access techniques

NASA in the U.S. and the Department of Communications in Canada have each invited agencies and organizations to submit proposals for communications experiments. Use of the spacecraft transponder will be allocated to approved experimenters with both countries sharing the available time equally.

CANADIAN USER PROGRAM

In Canada, the Department of Communications also will supply ground terminals which will be shared by the approved Canadian experimenters. A summary of the characteristics of planned Canadian ground terminals is given in Table 3.

15

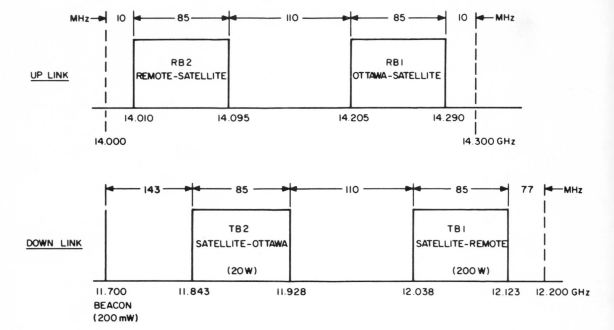

Fig. 6. SHF Frequency Plan

TABLE 3. Planned Canadian SHF Terminal Characteristics (12–14 GHz)

Terminal	Properties							
	Antenna		Receiver			Figure of Merit (G/T, dB/K)	Maximum Transmitter Power (watts)	Antenna Control
	Minimum Diameter	Peak Gain (dB) (12 GHz)	3 dB Beamwidth (degrees)	Type	System Noise Temperature (K)			
Control (Ottawa)	10 m (30 ft)	58	0.20	Uncooled paramp	425	32.8	Up to 300	Auto-track
TV remote transmission	3 m (10 ft)	49	0.53	Uncooled paramp	425	19.5	Up to 1000	Step-track
TV receive only two-way voice	2.4 m (8 ft)	47	0.70	TDA	1150	16.5	20	Manual adjustment
Two-way voice	1 m (3 ft)	37	1.84	TDA	1150	7.8	20	Non-tracking

The television broadcast to remote communities can be provided from the Control Terminal or the TV Remote Transmission Terminal. The receiving terminals will have an antenna with a diameter of 2.4 meter (8 feet) to permit limited multi-carrier operation with suitable back-off of the transponder power. With 100 W of transponder power, the terminal will provide television video with a signal-to-noise ratio of 46 dB (peak-to-peak video to weighted rms noise) and TV audio with a signal-to-noise ratio of 50 dB (test tone-to-noise). A two-

way voice channel will provide a telephone channel for use by remote communities or a voice/data channel for interactive ETV.

The TV Remote Transmission Terminal will be transportable by road, rail and (to some extent) air. It will use a 3-m (10-foot) diameter antenna to transmit either a network-quality picture from a remote location back to the main 10-m (33-foot) diameter antenna terminal at Ottawa or a lower-quality picture directly to communities. The prime requirement for this terminal is to transmit

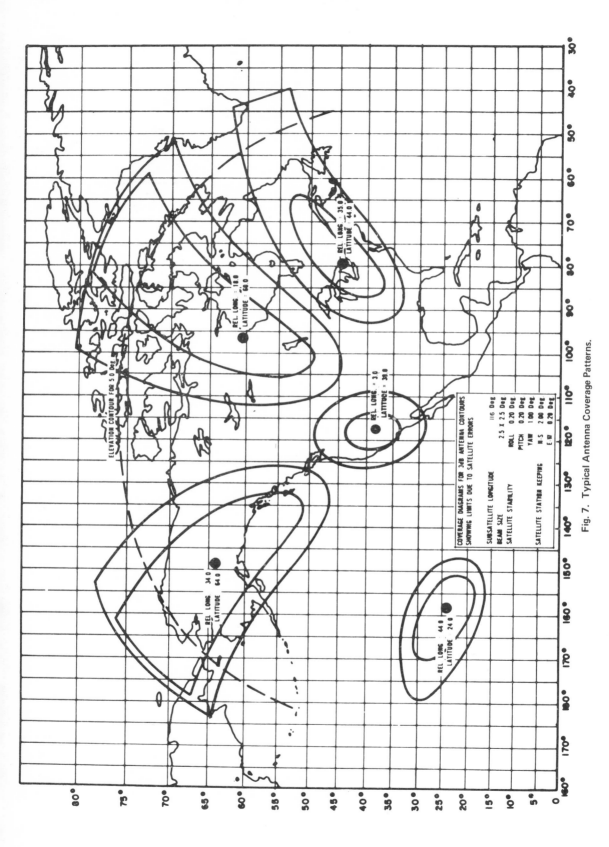

Fig. 7. Typical Antenna Coverage Patterns.

17

programs from areas not normally equipped with transmission facilities for live television.

For two-way voice, transportable terminals with 1-m (3-foot) diameter antennas are planned. These terminals will provide a two-way voice link, of telephone quality, between the 10-m (33-foot) terminal at Ottawa, or the 3-m (10-foot) transportable terminals, and isolated communities or survey parties. Present plans call for a single-channel-per-carrier FM system, with compandors, although digital delta modulation could also be used. Transmissions between 1-m (3-foot) terminals will require a double-hop circuit through the Ottawa or Transportable Terminal and will have the corresponding two-hop time delay.

At the end of 1974, the Department of Communications had tentatively approved 37 experiments.

U.S. USER PROGRAM

In the U.S., approved experimenters will provide their own ground terminals. NASA will provide technical consultative support in developing these ground terminals. As of June, 1975, seventeen U.S. experimenters have received conditional acceptance and several other proposed experiments are under evaluation. The experiments include health education, emergency communication, education in remote areas, and advanced communications technical experimentation.

The SYMPHONIE Project

DR. MARCEL R.E. BICHARA
Vibro-Meter AG
Fribourg, Switzerland

ACKNOWLEDGEMENTS

Acknowledgement is made to the Centre National d'Etudes Spatiales (Bretigny), the Fernmeldetechnisches Zentralamt (Darmstadt), Eurosat (Geneva) and in particular to my ex-colleagues of the Centre National d'Etudes des Telecommunications (Paris) for assistance and information given.

INTRODUCTION

In 1967 France and West Germany, having decided to cooperate on a project called SYMPHONIE, signed an intergovernmental Convention for the construction, launching and utilization of an experimental telecommunications satellite and for the design and construction of the terrestrial experimental stations needed for this project. Belgium, while not a formal participant, is involved in this project through its industries.

Since 1967 organizations at the governmental and industrial level have been created for studying, developing and manufacturing the various elements of the SYMPHONIE system. It is expected that the system will be operational by 1975 and will have a minimum lifespan of 5 years. This lifespan is determined by the reserves of propellant needed for maintaining the position of the satellite.

LAUNCH AND ORBITAL POSITION OF THE SATELLITE

The SYMPHONIE* system is of the geostationary type with the satellite maintained in a circular orbit around the equator at an altitude of about 36,000 km (22,000 miles) and stationed over the Atlantic Ocean at 11.5° W longitude. The nearly-circular orbit is brought to circularity by the action of motors with "hot gases" monomethylhydrazine, and nitrogen peroxide, N_2O_2.

The following operations then take place:
- The spinning rotation of the satellite is decreased by a "yoyo" arrangement;
- the panels of solar cells are deployed;
- a flywheel begins to spin rapidly around an axis parallel to the previous spinning rotational axis of the satellite;
- aided by a static infrared sensor, the satellite points to the center of the Earth;
- the rotation of the satellite on its own axis is stopped;
- the telecommunications antennas are deployed and permanently pointed to Earth. (The satellite is held in this position by means of a feedback loop comprising the infrared sensor and the spinning flywheel;
- the satellite is then flipped by 90° in such a manner that its orientation is north-south while the antennas remain pointed to the Earth.

The satellite will be maintained in position within ±0.5° in longitude and latitude (with a possibility of better than ±0.1°). The precision of pointing towards the Earth will be about ±0.5°.

SATELLITE TRACKING

Satellite tracking will be accomplished either by interferometry measurement of telemetry signals by two principal stations or by angular measurement of telemetry carrier signal direction. Interferometry measurement, if selected, will be conducted during the transfer and final orbit by station DIANE in Kourou, French Guyana, and station ZDBS at Weilheim, Germany. Measurement of telemetry carrier direction will be done by the SHF control station CNES, at Toulouse, France.

---

*One SYMPHONIE was launched on December 20, 1974 by a Delta 2914 launch vehicle.

TELECOMMUNICATIONS SYSTEM

The system comprises a space segment and an earth segment.

The Earth Segment

The earth segment includes:

(1) Stations with a figure of merit, G/T, of 31.5 dB/K. These will have a 15.5-m (50-foot) diameter auto-tracking antenna (58-dB gain; aperture efficiency 67%), two 3-kW transmitters, and a non-cooled parametric amplifier for reception. Presently, within the Convention framework, two stations are under construction, one in Germany at Raisting and the other in France at Pleumeur-Bodou. A third, belonging to the French Postal Authorities (P & T), will be built in the Reunion Islands.

(2) A mobile station of the French broadcasting authority (ORTF) with a G/T of 31.5 dB/K using a 12-m (40-foot) diameter antenna. This station will include a transmitter of several kilowatts, a cooled paramp and an automatic tracking antenna.

(3) Other stations to be erected by France with G/T of between 16.5 dB/K and 20 dB/K, using 3-to 4-m (10- to 13-foot) diameter antennas. These stations will have 10- to 50-W transmitters and low-noise receivers. Because of the larger antenna beamwidths, the antennas will be manually steered and will not need auto-tracking facilities.

(4) Other stations to be constructed by 1974-1975.

(5) Educational and data-transmission stations being planned.

The Space Segment

The telecommunications subsystem of the SYMPHONIE satellite will receive signals from and transmit signals to Earth coverage zones designated "Euroafrican" and "American." The subsystem will comprise a horn antenna for reception covering the whole of the zone of visibility; two parabolic transmitting antennas, each oriented towards its transmitting zone, and two transponders permanently connected to the receiving horn but connected to only one transmitting antenna at a time.

SUMMARY OF TECHNICAL DETAILS

A summary of the technical details is given in Tables 4 through 8.

TABLE 4. Frequencies*

	Earth to satellite	Satellite to earth
a	5940-6030 MHz	3715-3805 MHz
b	6065-6155 MHz	3840-3930 MHz
c	6195-6285 MHz	3970-4060 MHz
d	6320-6410 MHz	4095-4185 MHz

*Each band (uplink and downlink) is divided into 90-MHz subchannels.

TABLE 5. Mechanical and Dynamical Characteristics

- Mass while in the transfer orbit (with the apogee motors full of propellant): 387 kg (851 lbs)
- Mass, on station: 218 kg (480 lbs)
- Dimensions:

Equipment container: diameter	1.70 m (5.5 feet)
height	0.50 m (1.7 feet)
Diameter with panels deployed	6.86 m (22.5 feet)

- Stabilization:
  Principle: Three axes, controlled by wheel inertia, and for certain maneuvers, by gas jets.

Momentum of the wheel:	20.0 Nms
Attitude sensing elements:	a) sun sensor
	b) infrared static sensor
	c) "pencil-beam" infrared sensor

TABLE 6. Characteristics of Propelling Subsystem Components

Propelling subsystems	Apogee motor	Attitude control	Orbit control
Propellants	Aerozine 50 and nitrogen tetraoxide (N_2O_4)	N_2 under pressure	Mono-methyl-hydrazine and nitrogen tetraoxide (MMH/N_2O_4)
Mass of propellant	150 kg (300 lbs)	2.3 kg (5 lbs)	22.3 kg (45 lbs)
Specific impulse	303 s	68 s	285 s
Velocity increment	1440 m/s	–	381 m/s
Total impulse	415000 Ns	1220 Ns	84300 Ns
Number of jets	1	8	7
Thrust per jet	392 N	1 N	12 to 8 N ("blow-down" system)

*Aerozine 50 is a mixture of 50% hydrazine (N_2H_4) and 50% unsymmetrical dimethylhydrazine (UDMH).

TABLE 7. Electrical Power

Solar generator	
number of cells	21,888
semiconductor	silicon n/p
size	2 cm $\times$ 2 cm (0.79 in $\times$ 0.79 in)
panels	3 (cells on both sides)
power at end of life	167 W
voltage (regulated)	+ 27 V (with 12 other floating voltages)
Batteries (for eclipse operation)	2 with 14 Ni-Cd cells

TABLE 8. Telecommunications

Telemetering	
frequencies	VHF and SHF
telecommand	Standard NASA
	tone digital command system (TDCS)
	140 orders plus a system of teledisplay
Telemeasuring	Modulation by PCM/PSK/PM
	110 analog channels; 140 numerical
	channels; 5 series-numerical channels;
	rate: 8 and 64 bit/s
	VHF power: 0.7 W or 5.2 W
Antennas	toroidal pattern, circular
	polarization
Number of repeaters per	
satellite	2
Type	double frequency conversion
Repeater bandwidth	90 MHz
Receiving characteristics:	
frequency range	6 GHz
coverage	global
antenna	1 horn
beam size	17.2°
G/T	−15 dB/K
flux density required for transponder	
saturation	either −77 dBW/m^2, −79 dBW/m^2 or
	−87 dBW/m^2 depending on switched
Transmitter characteristics	received gain setting.
frequency	4 GHz
total R F channels	4 (2 per satellite)
coverage (for each	2 zones
transponder either zone)	(a) Europe/Africa and
	(b) America
antennas	2 parabolic reflectors illuminated by
	offset horns
beam size	13° × 8°
effective isotropically radiated power (EIRP)	29 dBW

The International Telecommunications Satellite Organization (INTELSAT)

MARTIN P. BROWN, JR
Communications Satellite Corporation
Washington, D.C., U.S.A.

In April 1965 INTELSAT initiated international satellite communications with the launch of INTELSAT I (Early Bird). Since that time, the system has expanded from a partnership of 14 nations to the present day consortium of 89. During this relatively short time period, traffic has increased from 132 voice-grade channels to well over 10,000, with the Atlantic region expanding at a rate of approximately 25% per year.

Global coverage is obtained through the use of spin stabilized, geostationary satellites, which can each view approximately 40% of the earth's surface. Accordingly, operational INTELSAT IV satellites are presently positioned over each of the major ocean regions at the following nominal locations (east longitude):

Indian Ocean — 61°
Pacific Ocean — 174°
Atlantic Ocean — 330.5° and 335.5°

The two Atlantic satellites carry about 60% of the total global satellite traffic. In addition to the operational satellites, each region has one spare satellite in orbit.

To keep up with the rapid traffic growth, a new satellite series has been introduced almost every two years; beginning with INTELSAT I (Early Bird) in 1965, INTELSAT II in 1967, INTELSAT III in 1968, and INTELSAT IV in 1971 (see Figure 8). The current series, as described in Table 9, will be augmented in the Atlantic region by the next series, INTELSAT IV-A, in the 1975-76 time frame. Nominal capacity of the INTELSAT IV is 7,500 voice-grade channels, and that of IV-A is about 12,500 voice-grade channels (plus TV and SPADE).

Earth station antennas used with the satellite network are typically 28- to 30-m (91.8- to 98.4-ft) diameter parabolic reflectors with a G/T of 40.7 dB/K. In recent years many countries have followed a trend toward construction of a second, and even a third, antenna to permit them to communicate with two satellites in one region or two different regions. There are currently 111 antennas at 88 earth stations in 64 countries. Additional information relating to standard INTELSAT IV earth station characteristics is shown in Table 10.

The types of modulated signals carried by the INTELSAT network include: data, facsimile, teletype, television (black and white or color), and voice. The primary modulation and satellite access technique used today consists of frequency-division multiplexed 4-kHz channels which frequency modulate carriers (FDM-FM) and are assigned to satellite transponders through frequency-division, multiple access (FDM-FM-FDMA). Other methods include single-channel-per-carrier (SCPC), pulse-code modulation/phase-shift key (PCM/PSK), and single-channel-per-carrier PCM multiple access demand-assigned equipment (SPADE). The majority of the traffic (80%), however, is transmitted through multicarrier transponder configurations using FDM-FM-FDMA.

In order to expand the capacity of the allocated satellite bandwidth in the 4- and 6- GHz bands, the new INTELSAT IV-A satellite will introduce the concept of frequency reuse through spatially separated antenna beams. Concepts under study for application with satellites which might be launched in the 1980 time frame are the use of both the 6/4-GHz and the recently allocated

Information furnished for this chapter is based on work performed under the sponsorship of the International Telecommunications Satellite Organization (INTELSAT). The views expressed are not necessarily those of INTELSAT.

TABLE 9. INTELSAT IV Satellite Parameters

PARAMETER	CAPABILITY
General	
No. satellites built	8 (F-1 thru F-8)
Design life	7 Years
Launch vehicle	Atlas (SLV-3C)/Centaur (D-1)
Apogee motor	Aerojet SVM-4A
Launch site	Eastern Test Range, Cape Kennedy, Florida
Launch management	NASA
Orbit	Geostationary (22,300 m), (35,800 km)
Major contractor	Hughes
Physical Description	
Stabilization	Dual spin (despun antenna earth oriented, spinning drum)
Cylinder diameter	238.1 cm (93.75 in.)
Cylinder height	281.9 cm (112.8 in.)
Height w/antenna mast	528.3 cm (208 in.)
Antenna configuration	
Global	2 Transmit / 2 Receive *
Spot†	1 West Transmit / 1 East Transmit
Telemetry	
During transfer orbit	1 dual mode, toroidal beam, bicone
Beacon #1 & #2	2 global transmit horns with 45° semi-reflectors
Command	1 dual mode, toroidal beam, bicone
Weight	
Launch (with adaptor)	1370.0 kg (3048 lbs.)
On station (initial)	696.0 kg (1547 lbs.)
Propellant	
Apogee AGC ANB-3066	637.0 kg (1416 lbs.)
Station keeping Hydrazine	122.0 kg (271.9 lbs.)

PARAMETER	CAPABILITY
Command	
Decoder	Redundant spinning & despun; 160 commands despun; 63 spinning
Frequency	6168 to 6182 GHz
Polarization	
Receive global	HCL
Transmit global	RHC
Transmit spot	RHC
Dual mode telemetry bicone	Linear
Transmit telemetry horn	RHC
Dual mode command bicone	Linear
Communication Capability	
Total assigned bandwidth	486 MHz
Total usable bandwidth	432 MHz
Assigned frequency range	
Uplink (receive)	5932-6418 MHz
Downlink (transmit)	3707-4193 MHz
Repeaters	
No. on-line	12
No. redundant	12
Type	Hughes 261-H TWTA
Bandwidth	36 MHz
RF power	6.3 W (8 dBW) nominal
Antenna gain (nominal)	
Global	
Beam edge‡	16.0 dBi
Beam center	19.5 dBi
Spot	
Beam edge	27.5 dBi
Beam center	30.5 dBi
Available EIRP (nominal)	
Global	
Beam edge	22.0 dBW
Beam center	25.5 dBW
Spot	
Beam edge	33.7 dBW
Beam center	37.0 dBW

Electrical Power	
Solar cells	2 × 2 cm, 45012 N/P with 12-mil cover glass
No. solar cells	42,240
Surface area	20.5 m²
Total DC power	569 W at 23.8 v (beginning of life, equinox)
No. batteries	2 (15 amp-hr, each, Ni-Cad)
Control system	
No. thrusters	
axial jets	2
radial jets	2
spin jets	2
Initial thrust per jet	25.5 N at 183 N/Cm³ (5.7 lbs at 265 psi)
No. inertial reference sensors	
Earth	2
Sun	2
Spin rate	45-75 rpm
Pointing accuracy (N-S, E-W)	
global	± 0.35°
spot	± 0.10°
Telemetry	
Encoders	Redundant spinning & despun; PCM; 64-8 bit words
Beacon #1	3947.25 MHz
Beacon #2	3952.50 MHz

Receive G/T (nominal)	
Beam edge	-18.6 dB/K
Beam center	-15.0 dB/K
Saturation flux density at beam edge	
Min. gain step attn. setting	-73.7 dBW/m²
Max. gain step attn. setting	-55.7 dBW/m²
Coverage	
Global or spot	Repeaters 1-8
Global only	Repeaters 9-12
Global beamwidth	17°
Spot beamwidth	4.5° (½ power)
Capacity (typical full configuration)	
Per global repeater	
FDMA/FM#	336-1092 channels
SPADE/TDMA	800 channels
1 TV	S/N = 54 dB
2 TV	S/N = 49 dB
Or combinations	
Per spot repeater	
FDMA/FM#	800-1872 channels
Total satellite capacity (typical)	
10 FDMA/FM repeaters	7500 channels
1 SPADE repeater	800 channels
1 TV repeater	TV

References:
– Astronautics and Aeronautics, June 1973
– COMSAT Technical Review, Volume 2, No. 2, Fall 1972

* One redundant
† Movable in 0.1° increments
‡ Beam edge = 8.5° from beam center for global, half power point for spot
# Telephony, telegraphy, facsimile, data, multicarrier or single carrier per transponder

INTELSAT I

Contractor: Hughes

1st Launch: 6 April 1965

Diameter: 72 cm (28.4 in.)

Height: 59 cm (23.3 in.)

Weight
 •Lift-Off: 67.5 kg (150 lbs.)
 •Orbit: 38.2 kg (85 lbs.)

No. Built: 2 { F-1 (Early Bird)
               F-2 (not launched)

INTELSAT II

Contractor: Hughes

1st Launch: 26 October 1966

Diameter: 142 cm (56 in.)

Height: 67 cm (26.5 in.)

Weight
 •Lift-Off: 161 kg (357 lbs.)
 •Orbit: 86.5 kg (192 lbs.)

No. Built: 5 { F-1 to F-4
               F-5 (not launched)

INTELSAT III

Contractor: TRW

1st Launch: 18 September 1968

Diameter: 142 cm (56 in.)

Height: 198 cm (78 in.)

Weight
 •Lift-Off: 284 kg (632 lbs.)
 •Orbit: 145 kg (322 lbs.)

No. Built: 8 (F-1 to F-8)

INTELSAT IV

Contractor: Hughes

1st Launch: 25 January 1971

Diameter: 238 cm (93.5 in.)

Height: 282 cm (111 in.)

Weight
 •Lift-Off: 1370 kg (3048 lbs.)
 •Orbit: 696 kg (1547 lbs.)

No. Built: 8 (F-1 to F-8)

Fig. 8. INTELSAT Series Spacecraft

TABLE 10. INTELSAT IV Earth Station Parameters (Standard Station)

PARAMETER	CAPABILITY
Antenna diameter	28-30 m (91.8-98.4 ft)
Gain-to-noise temperature ratio	
at 4 GHz	$\geqslant$ 40.7 dB/K
at other frequencies within 3.705-4.195 GHz band	$\geqslant$ 40.7 dB/K + $20 \log_{10} f/4$*
Antenna receive gain	
at 4 GHz	$\geqslant$ 57 dB
at other frequencies within 3.705-4.195 GHz band	$\geqslant$ 57 dB + $20 \log_{10} f/4$*
Transmit antenna sidelobes ($>$ 1° away from the main lobe center)	$\geqslant$ 29 dB below the main lobe max.
Transmitter size	1-12 kW wideband high power amplifier (500 MHz or tunable 1.8 kW klystrons (40 MHz))
Typical EIRP	75-95 dBW
Min. transmitter output backoff feed	−7 dB
Transmit polarization	LHC
Receive polarization	RHC
Axial ratio	1.4:1 (3 dB)
Antenna steerability	Compatible with satellites having $\leqslant$ 5° orbit inclination, ± 10° longitudinal drift
System bandwidth	
Receive	3.7-4.2 GHz
Transmit	5.925-6.425 GHz
Required EIRP stability	± 0.5 dB
RF out-of-band emission	
Spurious	$<$ 4 dBW/4 kHz
Intermodulation	$<$ 26 dBW/4 kHz

PARAMETER	CAPABILITY
Carrier frequency stability	
FM carriers above 5 MHz	± 150 kHz
FM carriers at or below 5 MHz	± 80 kHz
SPADE and PCM/PSK single channel-per-carrier preassigned carriers	± 200 Hz
RF energy dispersal or telephony carriers	Low-frequency triangular dispersal wave form calculated so that the maximum EIRP per 4 kHz of the fully loaded carrier is $\leqslant$ 2 dB.
Pre-emphasis for telephony, television, and program sound channels	In accordance with CCIR and CCITT recommendations
Typical modes of transmission Present:	1. FDM/FM with top baseband frequencies from 108 to 8,120 kHz; 2. FM for video basebands, including both 625 line/50 frame/s and 525 line/60 frame/s transmission; 3. PCM/PSK channels available on either a demand assignment basis (SPADE) or preassigned basis; 4. Digital/PSK channels for transmission of data, e.g. 50 kbit/s
Future:	5. TDMA 6. DITEC (digital TV)
Typical earth station-to-earth station service continuity	99.9%

*f is the receive frequency in GHz

References:
– ICSC-45-13 (Rev. 1) Performance Characteristics of Earth Stations in the INTELSAT IV System, 23 June 1972
– COMSAT Technical Review, Volume 2, No. 2, Fall 1972

14/11-GHz bands, satellite switching, and time-division multiple access (TDMA). New techniques under consideration for introduction into the INTELSAT system include: video phone, data phone, digital voice/data communications (DICOM), and digital TV (DITEC).

INTELSAT leases service to its members (or duly authorized communication entities), who may use the service themselves or lease it to other users, such as countries without earth stations. For the ultimate user, the cost of communication via satellite also includes charges for the earth station, land line to the gateway city of each country, and interconnection and switching; all of which are determined separately by rates usually established by the common carriers or operating entities in each country.

The cost of satellite use for INTELSAT members has been progressively reduced over the last several years. For example, the charge for the space segment per half circuit was $32,000/year in 1965, and $9,000 in 1974. Individual countries must defray all costs associated with their own earth stations and interconnecting links to their domestic terrestrial systems. Thus, the charge to the ultimate user is based upon both the space segment cost and the terrestrial system cost, as well as certain other related costs. Nevertheless, the cost to the commercial user has continued to decline.

The USSR Domestic System (MOLNIYA/ORBITA)

MARTIN P. BROWN, JR.
Communications Satellite Corporation
Washington, D.C., U.S.A.

The Soviet domestic satellite communication network can be described as follows (1–8):

Satellite Series . . . . MOLNIYA* I and II, STATSIONAR
Earth Stations . . . . ORBITA I and II

In accordance with directives of the 23rd Congress of the Communist Party of the Soviet Union, a system of television distribution from MOLNIYA I satellites to 20 ORBITA I earth stations was completed in 1967. Directives of the 24th Congress provided for further development and expansion to MOLNIYA II/ORBITA II equipment in the five-year period 1971–75. The first MOLNIYA II was launched November 24, 1971 and the first ORBITA II earth station inaugurated in September, 1972.

Basically, the MOLNIYA II satellites are upgraded versions of their predecessor with expanded bandwidth capacity at 6/4 GHz and are compatible with the ORBITA I and ORBITA II earth stations. Conceptually, the MOLNIYA/ORBITA system employs a large satellite with high-power output in order to economize on a large number of small earth stations. The heavy satellites (relative to the time of their first launch in 1967 and compared to EARLY BIRD) were possible with the use of the A-2-e launch vehicle which placed the satellites into a highly inclined elliptical orbit over the USSR with a period of 12 hours. To provide 24-hour coverage, from two to three spacecraft are placed in a phased orbit. Two antennas at each site are needed to execute a synchronized handover as one satellite passes over the horizon and another emerges.

---

*MOLNIYA means "lightning" in Russian (a colloquial meaning is "news flash"). Also spelled MOLNIA or MOLNYA, Russian молния.

The MOLNIYA/ORBITA system supplements local television broadcast systems with "Central Television" programs transmitted from two primary stations located in Moscow and Vladivostok. In addition to one television channel, some capability is provided for multichannel telephony, telegraphy, photofacsimile, and possibly digital communications. ORBITA I earth stations are receive-only systems, whereas ORBITA II stations may have the capability to transmit and receive.

As of May 1974, there were reported to be 51 ORBITA stations providing television broadcast to approximately 75% of the Soviet population. There are two ORBITA stations outside of the USSR—one in Ulan Bator, capital of the Mongolian Peoples' Republic, and one in Cuba.

On March 26, 1974, the USSR launched its first geostationary satellite known as STATSIONAR I (COSMOS 637). Indications are that COSMOS 637 was designed to study problems of launching into synchronous orbit and stabilization. On July 29, 1974, the Soviet Union launched its second geostationary communications satellite. Designated MOLNIYA A-1S, the spacecraft is in a circular orbit 35,850 km (22,000 miles) high with a period of 23 hours, 59 minutes and an inclination of 0 deg, 4 min (1). MOLNIYA A-1S is in use for experimental television and radio broadcasting. It is located over the equator at about 90° E longitude, southeast of Sri Lanka. It is believed that MOLNIYA A-1S is the first INTERSPUTNIK satellite which would serve as the beginning of a domestic and international satellite system with a global communications capability.

On September 30, 1971, an agreement was signed between the U.S. and USSR to replace the existing transatlantic cable and HF "Hot Line" with a redundant satellite network. Two systems will be operated, one between stations at Ft. Det-

PARAMETER	ORBITA I	ORBITA II
General		
Date of first site	November 1967	September 1972 (Arkhangel'sk)
Objective (No. sites/yr.)	6–8	
No. antenna per site	2 (handover)	Same
No. stations (total)		
USSR	48†	Unknown
Ulan Bator (Mongolian Peoples Rep.)	1	
Cuba		1
U.S. (Hot Line)		1
Major distribution stations	Moscow	Moscow
	Vladivostok	Vladivostok
Physical Features		
Antenna diameter	12 m (40 feet)	12 to 25 m (40 to 80 feet)
Type	Parabolic reflector	Same
Mount	Elevation over azimuth	Same
Manning	8 personnel	Same
Max. wind loading	180 km/h (110 mi/h)	Same
Temperature range	–48 to +46°C (–55 to + 115° F)	Same
Low-noise amplifier	Liquid nitrogen solid	Helium cooled
Communication Characteristics		
Mode	Receive only‡	Transmit and receive
Frequency range		
TV (up/down)	3700–3900 MHz	6185–6225/3860–3900 MHz
Telephony (up/down)	800–1000 MHz	Unknown/800–1000 MHz and
		5725–6169/3400–3844 MHz
Polarization	Circular	Same
Focal length	3 m (10 feet)	Same
Low-noise amplifier	Cooled	Same
TV parameters (color)		
Peak freq. deviation	15 MHz	Same
Upper modulating freq.	6 MHz	Same
System	French SECAM III (625 line/	Same
	frame, 25 frames/sec)	
Mode	FM	Same
IF frequency	70 MHz	Same
Sat. group delay compensation	Linear and parabolic	Same
Transmit power		5–10 kW
Tracking Characteristics		
General	Active reflector step track	Same
Method (normal)	Programmed by digital	Same
	computer from central	
	control at Shabalovka	
Drive (max.)	12°/sec. horizon to horizon	Same
Actual movement	Slow, due to high orbit	Same
Remarks		
Use	Domestic system	Domestic & international
		(INTERSPUTNIK)
Spacecraft	MOLNIYA I and II	MOLNIYA I & II & STATSIONAR

*Information shown in this table has been obtained from a variety of sources, and may require refinement when new (more complete) material is available.

†Some of which may be ORBITA II.

‡Except for Moscow and Vladivostok, which are both transmit and receive, and probably 25-m (40-foot) antennas, 5 kw.

#Pulse-deviation modulation on the back porch of the video line blanking pulses.

rick, Maryland (U.S.A.), and Vladimir (USSR), using MOLNIYA II satellites and another link between stations at Etam, West Virginia (U.S.A.), and a station near Moscow via an INTELSAT IV satellite. The Hot Line system between Etam and Moscow became operational in late 1974.

Characteristics of Soviet communications satellite earth stations are shown in Table 11, and those of communications satellites are shown in Table 12.

TABLE 12. Soviet Communication Satellite Parameters*
(Used with ORBITA Earth Stations)

PARAMETER	MOLNIYA I	MOLNIYA II	STATSIONAR
General			
First launch	23 April 1965	24 November 1971	26 March 1974
No. S/C launched	27†	6‡	2
(as of May '74)			
Launch site	Tyuratam/Plesetsk	Tyuratam/Plesetsk	Tyuratam
Launch vehicle	A-2-e#	A-2-e#	SL-12§
Spacecraft (physical)			
Height	3.8 m (11.3 feet)	3.8 m (11.3 feet)	
Diameter	1.7 m (5.2 feet)	1.7 m (5.2 feet)	
Weight	820–1000 kg	820–1000 kg	
	(1800–2200 lbs)	(1800–2200+ lbs)	
Life time (approx.)	2 years	2 years	
Positioning fuel	Nitrogen	Nitrogen (extra containers)	
Stabilization	Body	Body	
Power	500–700 W	700+ W	
No. silicon solar panels	12	24	
Sensors	Sun and Earth	Sun and Earth	
Shape	Pressurized cylinder	Pressurized cylinder	
Comm. capability			
Power (rf) TV	20/40 W¶	80 W	80 W
Telephony/data	7 W	7 W	7 W
No. TWT (active)			
80 W	–	1	1
20/40 W	3 (2 spare)	–	–
7 W	Unknown	20	20
TWT bandwidth			
80 W	–	40 MHz	40 MHz
20/40 W	50 MHz	–	–
7 W	Unknown	10 MHz	10 MHz
Total bandwidth			
(with guardband)	150 MHz+	500 MHz	50 MHz
Frequency range			
Up-link (TV)	6025–6225 MHz	6185–6225 MHz	6185–6225 MHz
Down-link (TV)	3700–3900 MHz	3860–3900 MHz	3860–3900 MHz
Up-link (telephony)	Unknown	5725–6169 MHz	5732–6176 MHz
		& unknown	
Down-link (telephony)	800–1000 MHz	3400–3844 MHz &	3407–3851 MHz
		800–1000 MHz	
Antenna			
Coverage	Global/Spot	Global	Global (Eurasia,
			Africa, Australia)
No. reflectors	2 (spot)	Global horns	
(one redundant)			

PARAMETER	MOLNIYA I	MOLNIYA II	STATSIONAR
Gain (at 1 GHz)	18 dB	Same	
Type	Parabolic	Same	
Accuracy	3° at apogee	Same	
Diameter	1 m (3 feet)	Same	
Telecommunications			
No. TV channels	3 (2 spare)	1	1
(B&W or color)			
Audio	Multiplexed with video	Same	Same
Other	Multiplexed telephony, telegraph, & photofax (FDMA-FDM)	Same	
Telephony quality		CCIR	CCIR
Orbit			
Inclination	65°	65°	Equatorial
Orbital period	12 hrs	12 hrs	24 hrs, geostationary
Apogee (Northern Hemisphere over USSR)	40,000 km (25,000 s. mi)	40,000 km (25,000 s. mi)	35,800 km (22,300 s. mi)
Perigee (Southern Hemisphere)	480 km (300 s. mi)	480 km (300 s. mi)	35,800 km (22,300 s.mi)
Position	Not stationary	Not stationary	1. 75–85° E longitude (Indian Ocean) 2. Over New Guinea
Remarks			
Purpose	Domestic telecommunications	Domestic telecommunications (upgrade of MOLNIYA I)	INTERSPUTNIK (similar in concept to INTELSAT)

*Information shown in this table has been obtained from a variety of sources, and may require refinement when new (more complete) material is available.

†MOLNIYA I-A through I-Z (letters I, Q, S, V, X, and Y were not used).

‡MOLNIYA 2-A through F.

#Modified four stage version of Sapwood (SS-6) ICBM, usually piggyback launch.

§ "Proton" booster plus 2 stages = 4 stages, originally an ICBM.

¶ Two TWT's are probably used, one with 20 W for use at low elevation angles and another (40-W) which is switched in to replace it as the satellite approaches higher elevation angles.

Blank spaces indicate that information was not available.

References

1. "Active Communication—Satellite Systems," Report 207-2, vol. IV, Part 2, XII Plenary Assembly of CCIR, New Delhi, 1970 (Green Book), pp. 51–54 and Draft Report 207-2 (Rev. 72), same subject, Conclusions of Interim Meeting of Study Group 4, Part 1, Geneva, 5-21 July 1972.
2. "Soviet Space Programs, 1966–70," Staff Report for the Committee on Aeronautical and Space Sciences, U.S. Senate, 92nd Congress, 1st Session, Doc. No. 92-51, December 9, 1971.
3. "Soviet Space Programs, 1971," Staff Report for the Committee on Aeronautical and Space Sciences, U.S. Senate, 92nd Congress, 2nd Session, April 1972.
4. L. Ya Kautor, V.A. Polukhin, and N.V. Talyzeu; Moscow, *Elektrosvyaz*, no. 5, 1973, pp. 1–8, signed to press 3 January 1973, in Russian.
5. *TRW Space Log*, 1973.
6. "List of Space Radiocommunication Stations and Radio Astronomy Stations," ITU International Frequency Registration Board (IFRB), 4th Edition combined with Supplement No. 1, 1 November 1973.
7. "Intersputnik, International Space Communication System Organization," I. Petrov, *Telecommunication Journal*, vol. 39, pp. 679–682, November 1972.
8. "Synchronous MOLNIYA," *Aviation Week and Space Technology*, August 5, 1974.

The Canadian Domestic System (TELESAT)

ROBERT K. KWAN
TELESAT Canada
Ottawa, Canada

TELESAT—THE CANADIAN DOMESTIC SYSTEM (1–4)

Telesat Canada is a Canadian corporation, established by an Act of the Canadian Parliament, that came into being in September 1969. It is authorized to install and operate a multi-purpose system of communications by satellites throughout Canada. This system is capable of providing television, broadcast radio, data and facsimile transmission services throughout Canada, and can handle both analog and digital signals. It augments and links with the existing terrestrial facilities.

The space segment consists of three spin-stabilized satellites in geostationary orbit. These satellites are named ANIK, an Eskimo word for "brother." On November 9, 1972, ANIK I was launched and subsequently placed at 114°W longitude. Commercial telecommunication services began in January 1973. ANIK II, which was launched on April 20, 1973, and is located at 109°W, is functioning as a second operational satellite and an in-orbit backup. ANIK III, which will be launched in early 1975 and will be located at 104°W, will guard against premature failure of either of the first two satellites and will provide for ultimate system expansion.

The ground segment of the system (as of December 31, 1973) consists of six types of earth stations with 37 stations operational and an additional 15 being completed.

The following sections describe the spacecraft system, the satellite control system, and the earth stations. A summary of system applications is given for the provision of communications services.

Spacecraft System

The Telesat satellites are active spin-stabilized multi-channel repeater communications satellites for use in geostationary orbit. Each satellite has twelve RF channels—10 operational and 2 protective. The receive frequency of these channels is in the 5.925- to 6.425-GHz band; and the transmit frequency is in the 3.7- to 4.2-GHz band. Telemetry, tracking and command functions required for satellite station keeping and positioning are also provided in these frequency bands. Each RF channel in the communications subsystem of the spacecraft is essentially an independent transponder with a usable bandwidth of 36 MHz. Channels are separated by a 4-MHz guard band. Each channel can be operated with one or more carriers in either a saturated or linear mode. The only active equipment common to all channels is a redundant wide-band receiver which translates the 6-GHz carriers to 4 GHz, and amplifies the 4-GHz carriers to an intermediate power level prior to channelization. The major satellite characteristics are summarized in Table 13.

TABLE 13. Telesat Satellite Characteristics

Number of RF channels	12 (10 during eclipse)
RF channel spacing	4 MHz
EIRP per RF channel	33 dBW (minimum)
Receive G/T	−7 dB/K
TWT output power	7 dBW (5W)
Saturating power flux density per RF channel	−80 dBW/m^2
Antenna coverage	all Canada
Orbit inclination control	±0.1°
Longitude drift control	±0.1°
Design life	7 years
Frequency band	
Transmit	3700–4200 MHz
Receive	5925–6425 MHz

Satellite Control System

The satellites, once their geostationary positions are achieved, are kept on station by precise station-keeping maneuvers. These include orbit and attitude corrections to within ±0.1° of longitude, latitude and orientations.

The tracking, telemetry and command (TT&C) facilities required for normal operations are co-located with the Heavy Route station at Allan Park in Ontario. All TT&C activities are controlled from Telesat's Satellite Control Center in Ottawa. Control of the TT&C facilities is achieved by voice and data circuits. Computers located at the TT&C site and at the Satellite Control Center handle the data processing and the TT&C functions.

The Earth Segment

The earth-station configuration includes 37 stations which can be classified into six different types; namely, the Heavy Route (HR), Network TV (NTV), Northern Telecommunication (NTC), Remote TV (RTV), Thin Route (TR), and TT&C earth stations. The major characteristics are summarized in Table 14. With the exception of the HR and TT&C stations, the remaining stations are designed to operate without full-time personnel in attendance. Further, to maintain the station-keeping of the satellites to within ±0.1° in longitude and orbit inclination, manual tracking capability is provided at each station, except at the two HR stations where the antenna beamwidth is sufficiently narrow that automatic tracking is needed. The TT&C station is also equipped with full tracking capability because of its important role in placing and maintaining the satellite in its precise orbit.

Communications Services

As of December 31, 1973, eight of the ten RF channels are assigned to commercial service in the first satellite. A summary of these services is given in Table 15. As shown, three RF channels are assigned to the National Television Distribution Services in both the English and the French languages. This TV distribution network serves 35 locations and the service was commenced in April 1973. Table 16 summarizes the noise performance requirements in the TV service.

Five satellite RF channels are assigned to telephone message traffic. Two of these are operated in a single-carrier FM mode to provide a 960-channel voice capacity between the two HR stations. The noise performance for each voice channel is 37.5 dBrnc0 or 5625 pWp0.

A third channel is assigned to medium telephone message traffic between three locations (1 HR and 2 NTC stations) that require 12 to 60 circuits. In this application the FM/FDMA mode is used. Noise performance requirements are 37.5 dBrnc0 to be received at the southern HR station and 44 dBrnc0 to be received at the northern NTC stations.

A fourth channel is assigned to light telephone message service providing 2 to 8 circuits to small, remote communities in the Canadian far North. This single-voice-channel-per-carrier approach utilizes the delta modulation PSK/FDMA technique and is known as the Thin-Route Service.

A fifth channel is used to provide a 240-channel voice circuit (3 kHz) facility between an HR and an NTV station, using the conventional two-carrier FM/FDMA technique. The noise performance of 7200 pWp0 is predicted for this type of service. By September 1975 this two-access FDMA system will be replaced by a more efficient PCM/PSK/TDMA system. This TDMA system will provide an increase in circuit capacity to 400 circuits with an improved noise performance objective of better than 2700 pWp0. This system utilizes a 9-bit PCM codec and a 4-phase PSK modem operating at a channel bit rate of about 61 Mb/s. Detailed information on this TDMA system is given in Reference 5.

TABLE 14. Telesat Earth Station Characteristics

	HR	NTV	NTC	RTV	TR	TT & C
Number of stations	2	6	2	25	17	1
Antenna						
Diameter (ft)	97 (30.5m)	33 (10m)	33 (10m)	26 (8m)	26 (8m)	36 (11m)
G (6 GHz) (dB)	62	52	52	51	51	54
G/T (4 GHz) (dB/K)	37	28	28	26/22	22/19	28
Transmitters						
Installed/station	3–8	1–3	2	0	2	2
EIRP/channel (dBW)	83	83	73	0	53	85
Receivers						
Installed/station	5–10	4	2–3	1–2	2	2
Antenna steering	Step track	Manual	Manual	Manual	Manual	Monopulse
No-break standby power	Batteries & Diesel	Batteries & Diesel	Batteries & Diesel	Batteries	Batteries	Batteries & Diesel

TABLE 15. Canadian Domestic Communications Satellite System Applications

Tele-communication service	Number of Satellite RF Channels	Number of Locations Served	Communication Technique	Service Commencement Date
Television distribution	3	35	Video-FM Audio-FDM-FM	April 1973
Heavy route telephone message (east-west)	2	2	FDM-FM	January 1973
Medium route telephone message (north-south)	1	3	FDM-FM-FDMA	January 1973
Thin route telephone message (small remote communities)	1	2	Delta-PSK-FDMA	February 1973
		17		January 1975
Telephone message* (Toronto–CANTAT II cable)	1	2	FDM-FM-FDMA	April 1974
			FDM-PCM-PSK-TDMA	September 1975

*The initial FDM-FM-FDMA facilities will be replaced in 1975 by FDM-PCM-PSK-TDMA.

TABLE 16. Television Performance

	HR & NTV Stations	RTV Stations
Video S/N	54 dB*	52 dB*
1st & 2nd Audio S/N	56 dB	53 dB
Control Audio S/N	40 dB	40 dB

*Peak-to-peak picture signal to rms weighted noise.

References

1. R.F. Chinnick, "The Canadian Telecommunications Satellite System," *Journal of the British Interplanetary Society*, vol. 26, pp. 193–202, 1973.
2. J. Almond, "The Telsat Canada Domestic Communication Satellite System," presented at CCITT Latin America Region Plan Meeting, Brasilia, Brazil, June 25–July 6, 1973.
3. R.M. Lester, "Television Distribution by the Canadian Domestic Satellite System," *Journal of the SMPTE*, pp. 88, February 1972.
4. R.M. Lester, "Canadian Domestic Satellite System Applications," *IEEE Intercon 73*, New York, March 1973.
5. R.K. Kwan, "A TDMA Application in the Telesat Satellite System," Proc. *IEEE National Telecommunications Conference*, Atlanta, Georgia, pp. 31E-1–31E-6, November 1973.

US Domestic Communication Satellite Systems

RICHARD G. GOULD
Satellite Systems Engineering, Inc.
Washington, D.C., U.S.A.

The Federal Communications Commission has authorized companies to construct, own and operate domestic communications satellite systems in the 4- and 6-GHz bands. It also authorized two interim systems based on the lease of transponders in the ANIK satellites of Telesat Canada, in conjunction with earth stations in the United States. The status and general characteristics of these systems are shown in Table 17. Detailed system parameters are given in later sections. Originally, American Satellite Corporation (ASC) constructed three earth stations in the United States. ASC had authority to lease three transponders in the ANIK satellites to provide commercial communications service between these stations, but subsequently transferred that lease to the Western Union satellites. In addition, ASC has constructed five earth stations on or near Air Force installations which provide circuits linking these sites with Offutt AFB in Nebraska.

RCA Global Communications and RCA Alaska Communications also had authorization to lease transponders in the ANIK satellites. These transponders were used with several earth stations owned and operated by RCA in Alaska and in the 48 contiguous states. More recently, RCA transferred its lease to the Western Union satellites. In addition, RCA has a pending application for a 24-transponder, body-stabilized spacecraft of its own design and manufacture and is now constructing such spacecraft in preparation for an early 1976 launch.

Western Union has launched two 12-transponder spacecraft built by the Hughes Aircraft Company.

These satellites provide links between the several earth stations owned and operated by Western Union in the United States.

The Commission has also authorized a system, comprising satellites to be built by the Hughes Aircraft Company, which will be owned and operated by Comsat General Corporation, a wholly-owned subsidiary of the Communications Satellite Corporation, and earth stations to be owned and operated by the American Telephone and Telegraph Company (AT&T). GTE Satellite Corporation, a wholly-owned subsidiary of General Telephone and Electric Company, received Commission approval for a separate system, but recently applied to the Commission to abandon its plans for the lease of its own satellite capacity from National Satellite Services, Inc., a wholly-owned subsidiary of the Hughes Aircraft Company, and instead to participate in the Comsat General/AT&T system. The new joint system would have a total of seven earth stations—four of AT&T and three of GTE, plus 2 TT&C stations.

In addition to the earth stations mentioned in connection with each of the systems discussed, many more applications for earth stations are on file, but have not been acted on at the request of the applicant. Such station applications have been frequency-coordinated to prevent subsequent applicants for terrestrial stations from applying for frequencies or facilities that would cause them interference. Applicants for earth stations may request that permits be issued for one or more of these stations as traffic requirements and business conditions permit.

Many of the applications are for receive-only stations that would receive TV programming for subsequent distribution through a CATV system

---

Information contained in this section is derived from information on file at the FCC, available to the public.

TABLE 17. Domestic Satellite Systems

Company	Earth Station Locations	Services	Satellites	Status
Western Union	New York Chicago Los Angeles Dallas Atlanta	Data, voice and video leased private line	Two 12-channel HS-333's (WESTAR) satellites built by Hughes Aircraft Company	Granted. (operational July, 1974)
American Satellite (A.S.C.)	New York Los Angeles Dallas	Data, voice and video leased private line	Phase 1: Lease of 3 channels in WESTAR	Granted. (operational July, 1974)
	Fairchild AFB, Wash. Loring AFB, Maine Centerville Beach, Calif. Moffett Field, Calif. Offutt AFB, Nebr.		Phase 2: 12-channel HS-333's	Postponed by A.S.C.
RCA Global Commun. RCA Alaska Commun. (RCA)	New York Los Angeles San Francisco Juneau, Alaska Fairbanks/Anchorage	Data, voice and video leased private line plus MTT within Alaska and between Alaska and CONUS	Phase 1: Lease of 2 channels in ANIK (or WESTAR) Phase 2: Two 24-channel satellites built by RCA Astro-Electronics	Granted. (operational Jan. 1974) Pending (scheduled for launch Jan. 1974)
AT&T	New York Chicago San Francisco Atlanta Los Angeles*	MTT (no private line except to the U.S. government for three years)	Leased from Comsat General by AT&T	Earth station authorizations granted.
GTE Satellite	Los Angeles Tampa Honolulu	MTT (no private line except to the U.S. government for three years)	Leased from Comsat General by AT&T	Earth station authorizations granted. GTE application to participate in joint system with AT&T; decision pending
Comsat General	TT&C only (Santa Paula, Calif. and Southbury, Conn.)	Lease of transponders to AT&T	Three 24-channel satellites (mod. of INTELSAT IV) built by Hughes Aircraft Company	Lease to AT&T granted.
General Electric Co.	Valley Forge, Pa. Daytona Beach, Fla.	Company communications; Equipment and system development	Transponders leased from Western Union	Operational

*If the GTE Satellite application to participate in a joint system is granted, as discussed in the text, the Los Angeles earth station will not be built by AT&T.

or for rebroadcast through conventional, existing TV broadcast stations.

In addition to the authorized stations mentioned in Table 17, there have been several temporary, experimental or demonstration stations authorized from time to time. For example, the Teleprompter Corporation operated a transportable receiver-only station with an 8-m (25-foot) diameter antenna at many locations around the country to assess and demonstrate the feasibility of TV reception for cable system use. (The Telesat ANIK satellite was used and provided useful signals even in the southern states although its antenna pattern had been shaped to provide nearly uniform coverage of Canada.)

The General Electric Company has authorization for earth stations which are described later.

COMSAT GENERAL CORPORATION SATELLITES

General

The satellites of Comsat General Corporation will contain 24 transponders. Their design is similar in many respects to that of the INTELSAT IV.

TABLE 18. Satellite Specifications

Item	Description
Launch vehicle type	Atlas/Centaur
Satellite attitude stabilization east–west north–south	 ±0.20° ±0.26°
Stationkeeping	±0.1°
Orbital inclination	±0.1°
Power source	Solar array-storage batteries to permit full operation during eclipse periods
Antenna coverage	Continental U.S., Alaska, Hawaii, and Puerto Rico
Polarization	Linear (frequencies reused by horizontal and vertical cross-polarization).
Polarization isolation (receive or transmit)	 33 dB
Transponders per satellite	24
Transponder distribution Satellite receive	 CONUS and Alaska: 12 CONUS, Hawaii and Puerto Rico: 12
Satellite transmit	CONUS: 12 to 24 (switched) Alaska only or CONUS-Alaska: 6 to 0 (switched) Hawaii/Puerto Rico: 6 to 0 (switched)
Transmit frequency range	3700–4200 MHz
Receive frequency range	5925–6425 MHz
EIRP per transponder (within coverage areas*)	 33.0 dBW
G/T (within any coverage area)	–9 dB/K
Usable rf bandwidth	34 MHz per transponder
Useful life	7 years

*CONUS-Alaska combined coverage area is 31 dBW.

Satellite specifications are shown in Table 18 and the Frequency Plan is illustrated in Figure 9. A block diagram of the satellite communication system is shown in Figure 10.

The 24 transponders can provide coverage of the 48 contiguous United States (CONUS). The outputs of six of these transponders (as indicated in Figure 10) will be switchable from CONUS to transmit down-link signals to both Hawaii and Puerto Rico. Additionally, the outputs of six other transponders will be switchable from CONUS to transmit downlink signals to Alaska only or to a combined CONUS-Alaska service area. The inputs to twelve transponders will receive uplink signals from Alaska and CONUS. Additionally, the inputs to twelve other trans-ponders will receive uplink signals from CONUS, Hawaii and Puerto Rico.

Antennas

CONUS coverage: Two 1.5-m (60-inch) diameter reflectors with multiple feeds providing a beamwidth of approximately 7.0° × 3.2°.

Alaska/Hawaii/Puerto Rico coverage: Additional feeds located in the reflectors to produce beams of approximately 4.0° × 2.0° for Alaska, and 3.0° each for Hawaii and Puerto Rico.

By connecting the transponder outputs to both the CONUS and Alaska feeds, a combined coverage of CONUS and Alaska is achieved.

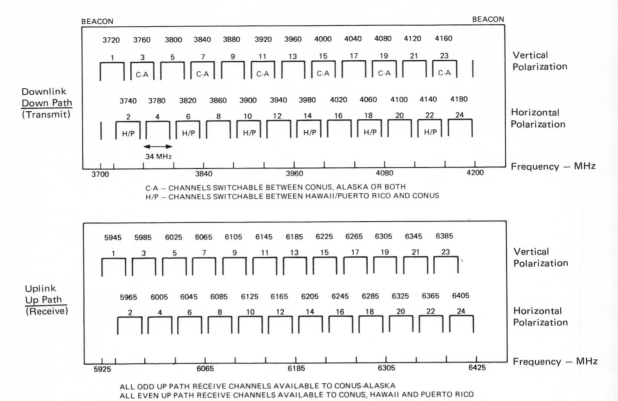

C-A — CHANNELS SWITCHABLE BETWEEN CONUS, ALASKA OR BOTH
H/P — CHANNELS SWITCHABLE BETWEEN HAWAII/PUERTO RICO AND CONUS

ALL ODD UP PATH RECEIVE CHANNELS AVAILABLE TO CONUS-ALASKA
ALL EVEN UP PATH RECEIVE CHANNELS AVAILABLE TO CONUS, HAWAII AND PUERTO RICO

Fig. 9. Frequency Plan for the COMSAT/AT&T Satellite

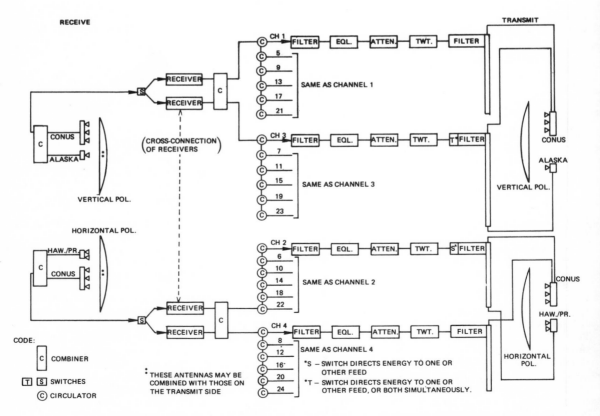

Fig. 10. Communication Subsystem Block Diagram of the COMSAT/AT&T Satellite

Telemetry: dual mode, toroidal beam, omnidirectional

Command: toroidal beam, omnidirectional

Equivalent Isotropically Radiated Power (EIRP)

CONUS coverage: 33 dBW
Alaska only, Hawaii, and Puerto Rico coverage: 33 dBW
CONUS-Alaska combined coverage: 31 dBW
Telemetry: 5 dBW

Flux Density

For a satellite EIRP of 36 dBW (which will be the maximum EIRP within the coverage areas), the flux density on the surface of the earth is calculated below:

Satellite maximum EIRP	36	dBW
Path loss (free space)	−196.5	dB
Gain of one-square-meter antenna (at 4 GHz)	33.5	dB
Flux density:	−127.0	dBW/m^2
Reduction of power flux density in a 4-kHz band due to energy dispersal	25	dB
CCIR requirements at 4 GHz and 0° elevation angle	−152	dBW/m^2/ 4 kHz

Carrier energy dispersal will be achieved by the modulating signal, and may require the addition of a spreading waveform in the transmissions from the earth stations to the satellites.

Launch Vehicle

Atlas/Centaur

Station-Keeping Functions

The satellite will be controlled to the following orbital and attitude parameters:

Orbital inclination	⩽0.1°
Spacecraft pointing error	
East–west	± 0.20°
North–south	± 0.26°
Longitudinal drift (station keeping)	⩽± 0.1°

MM-Wave Propagation Experiment

A millimeter-wave beacon operating in the 18- and 30-GHz band will be provided on each satellite.

AMERICAN TELEPHONE & TELEGRAPH COMPANY EARTH STATIONS

General

Originally, AT&T applied for, and was granted, authorization to build five earth stations. If the application of GTE Satellite Corporation to participate in the AT&T/Comsat General system is granted (as described above), AT&T would then build, own and operate only four stations: Three Peaks, California, near San Francisco; Woodbury, Georgia, near Atlanta; Hanover, Illinois, near Chicago; and Hawley, Pennsylvania, to serve the New York City area. Typical characteristics of the AT&T stations are given below. A functional block diagram is shown in Figure 11. TT&C stations owned and operated by Comsat General will be located at Santa Paula, Calif. and Southbury, Conn.

Frequencies

Transmission from the earth stations will be accomplished on frequencies in the 5925- to 6425-MHz band, and reception will take place in the 3700- to 4200-MHz band.

Communications Capacity Under Proposed System Operating Conditions

Telephone: 4800 voice circuits (1200 in each of four equipped radio-channel pairs)

Transmitters

1. Frequency range: 5925 to 6425 MHz
2. Frequency stability: 0.001% (crystal control)
3. Required bandwidth: 34 MHz per channel
4. Power output at antenna feed: 3000 W

Antennas

1. Type: Two steerable paraboloids with cassegrain feed.
2. Size: Diameter 30 m (100 feet)
3. Gain: 62.8 dB at 6 GHz (referred to transmitter side of diplexer)
   60.4 dB at 4 GHz (referred to receiver rf amplifier input)
4. Beamwidth: 0.12° at 6 GHz
   0.18° at 4 GHz
5. Sidelobe patterns: Sidelobes will meet or exceed the FCC antenna performance standard of Section 25.209: $G = 32 - 25 \log \theta$ dB above isotropic for off-beam angles of $1° < \theta < 48°$ and −10 dB with respect to isotropic for $\theta > 48°$.
6. Polarization: Linear; horizontal and vertical
7. Tracking mode: Automatic tracking
8. Receive antenna G/T = 41.4 dB/K at 4.0 GHz (minimum, at 20° elevation angle)
   G = 60.4 dB (adjusted by 0.5 dB for antenna losses)
   T = 70 K = 19.0 dBK
9. EIRP
   a. Main beam: 92.3 dBW (max) per channel
   b. Horizontal plane (in any 4-kHz band): 5.0 dBW (max) at an elevation of 20°.

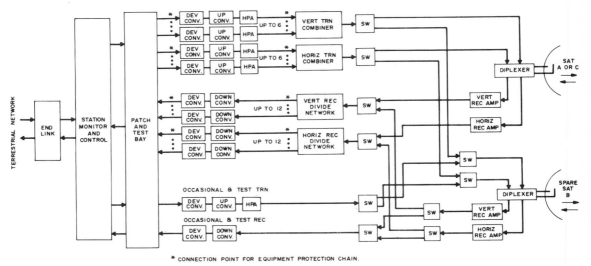

Fig. 11. AT&T Earth Station Block Diagram

* CONNECTION POINT FOR EQUIPMENT PROTECTION CHAIN.

GTE SATELLITE CORPORATION EARTH STATIONS

General

GTE originally applied for, and was granted authorization to build, five earth stations. Subsequently GTE has requested a modification of its authority that would result in its building, owning and operating only three stations: Triunfo Pass, California, near Los Angeles; Homosassa, Florida, near Tampa; and on Oahu near Honolulu, Hawaii. These stations would be operated in conjunction with four to be built, owned and operated by AT&T in a joint system using the same Comsat General Corporation satellites. Typical characteristics of the GTE stations are shown in the following. A functional block diagram is shown in Figures 12a-b.

Communications Transmitter and Antenna System

Transmitter:	
Frequency range	5925 to 6425 MHz
Bandwidth	40 MHz
Power Output:	
Klystron	3000 W, 35 dBW
At antenna input	1000 W, 30 dBW
Effective isotropically radiated power (EIRP)	92 dBW maximum
Maximum modulating frequency	6000 kHz
Maximum deviation	±15,000 kHz
Frequency tolerance	0.0001% (crystal controlled)
Radiation in horizontal plane	+3 dBW per 4 kHz (depends on earth station location and siting)

Antenna

Type	Cassegrain, fully steerable, 30-m
	(100-foot) diameter
Minimum elevation angle	20°
Antenna gain at 6 GHz	62.5 dB nominal
Beamwidth at 6 GHz	0.12° nominal
Polarization capability	Linear orthogonal with any orientation, and circular
Receiving system G/T	43 dB/K

WESTERN UNION TELEGRAPH COMPANY SATELLITES

General

The Western Union satellites are almost identical to the ANIK satellites built for Telesat Canada by the Hughes Aircraft Company. They have 12 transponders. Satellite specifications and characteristics are discussed in the following. Frequency plan is illustrated in Figure 13. The block diagram of the communications system is shown in Figure 14.

WESTERN UNION TELEGRAPH COMPANY EARTH STATIONS

General

Western Union has built, owns and operates five earth stations at: Glenwood, New Jersey, to serve New York City; Steele Valley, California, near Los Angeles; Estill Fork, Alabama, to serve Atlanta; Cedar Hill, Texas, to serve Dallas; and Lake Geneva, Wisconsin, to serve Chicago. Typical characteristics are given in the following.

Polarization

The antenna polarization is linear, rotatable at both 4 and 6 GHz.

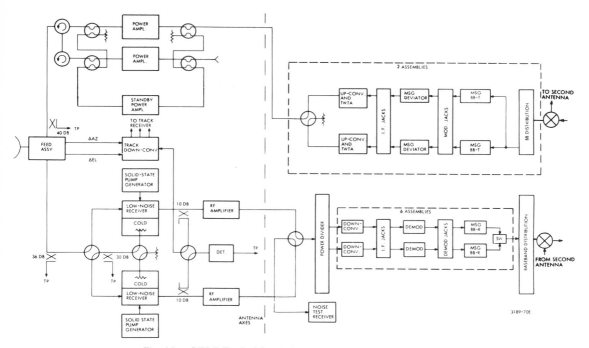

Fig. 12a. GT&E Typical Earth Station Functional Block Diagram

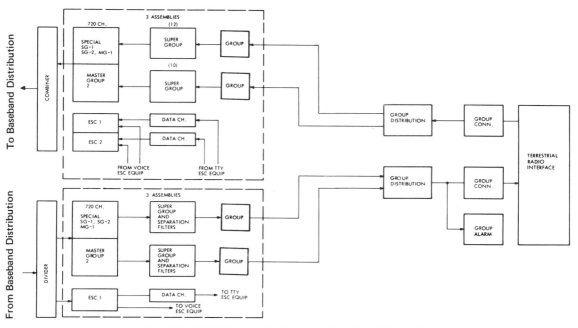

Fig. 12b. GT&E Typical Earth Station Functional Block Diagram

All transmissions to the twelve satellite transponders are vertically polarized; i.e., perpendicular to the satellite spin axis.

All telemetry signals from the satellite are polarized perpendicular to the satellite spin axis.

Maximum Modulating Frequency

5.26 MHz (with 6.314 MHz pilot)

Transmitters

1. One operating transmitter with one standby transmitter will be employed initially. The transmitters are Energy Systems Model 1032A or equivalent, equipped with 1.5 kW Varian klystrons (VA-936-J).

2. Frequency Range: 5.925–6.425 GHz
3. Frequency Stability: ±30 KHz in 24 hours
4. Frequency Accuracy: ±50 KHz

Satellite Characteristics

ELEMENT	DESCRIPTION
1. Satellite	Spin-stabilized, geostationary orbit
Station keeping accuracy	Better than ±0.1° overall
Attitude stability	Better than ±0.1°
Power source	Solar-cell array; nickel-cadmium battery for eclipse reserve; 220 W minimum prime power
Control system (orbit/attitude)	Hydrazine propulsion by thruster jets; ground command control; capable of restationing to new longitude
Weight	510 kg (1120 pounds) pad weight 250 kg (549 pounds) initial weight in orbit
Apogee motor	Solid propellant; velocity increment = 1840 m/s (6028 ft/s)
Design lifetime	7 years
2. Transponder	Channelized, linear heterodyne type; 12 channels optimized for single-carrier operation; BW/channel = 36 MHz; full 12-channel operation at beginning of design lifetime; 10-channel operation during eclipse and at end of design lifetime
Frequency (includes telemetry and command)	Transmit: 3.7−4.2 GHz Receive: 5.925−6.425 GHz
3. Communications antennas (also for telemetry and command when on station)	Mechanically de-spun 1.5 m (60-inch) parabolic reflector with multiple feeds; transmit and receive all channels; linear polarization; transmit perpendicular to spin axis; receive parallel to spin axis
Beamwidth	Contiguous U.S.: 6.8° × 3.5° (main beam) Alaska: 2.8° (pencil beam, squinted) Hawaii: 2.8° (pencil beam, squinted)
Pointing control	RF tracking (monopulse) of pilot signal from each station with command offset
Pointing accuracy	Better than ±0.1°
Sidelobes	−25 dB from main lobe
Telemetry	Dual mode, toroidal beam, omnidirectional; linear polarization; (Used for launch and transfer−orbit phase)

Spacecraft System Parameters

	Contiguous United States	Alaska/Hawaii
1. Antenna/transponder (overall)		
Receive G/T	−7.4 dB/K	−14.4 dB/K
Single-carrier EIRP	33.0 dBW	26.0 dBW
Single-carrier flux density required to saturate final TWT	−80.0 dBW/m^2	−73.0 dBW/m^2
2. Carrier IF bandwidth		
Single carrier/transponder	36.0 MHz	—
Two carriers/transponder	10.0 MHz	10.0 MHz

Multipoint Contiguous States Service (1200 Circuits/Channel)

Spacecraft EIRP/channel	33.0 dBW

	U.S. to Alaska/Hawaii	Alaska/Hawaii to U.S.
Spacecraft maximum EIRP/channel	33.0 dBW	33.0 dBW
Two-carrier loss	-5.0 dB	-5.0 dB
Alaska/Hawaii beam loss	-7.0 dB	0
EIRP/Carrier	21.0 dBW	28.0 dBW

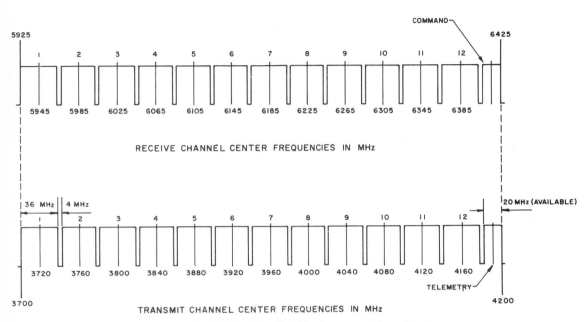

Fig. 13. Western Union Satellite Frequency Channelization

5. Necessary bandwidth: 36 MHz
6. Power output at antenna feed:
   25.6 dBW nominal (consistent with 83 dBW EIRP)
7. Emission limitations:
   a. All harmonics from klystron down 60 dB from carrier.
   b. EIRP beyond the 36 MHz operating channel but within the 5.925–6.425 GHz band resulting from spurious tones or other unwanted signals not to exceed 4 dBW in any 4 KHz band.

Antenna

1. The antenna subsystem is a 17-m (50-foot) diameter parabolic reflector with a cassegrain multimode horn feed mounted on a limited motion elevation-over-azimuth pedestal. A step-tracking system is employed.
2. Reflector f/D: 0.40
3. Reflector diameter: 17 m (50 feet)
4. Angular coverage: Elevation 10° to 59°
                             Azimuth ±56°
5. Gain
   4.00 GHz: 54.7 dB
   6.00 GHz: 57.7 dB

6. G/T system referred to LNA input, 30° elevation
   4.00 GHz: 37.4 dB

7. The first sidelobes are to be at least 14 dB below the main lobe maximum of the transmit pattern. Sidelobes at 1° or more away from the main lobe center are not to exceed a gain of 32–25 log θ dB above isotropic, where θ is the angle e from the main lobe in degrees between 1° and 48°. For angles between 48° and 180° away from the main lobe center, the sidelobes are not to exceed a gain over isotropic of -10 dB.

RCA GLOBAL COMMUNICATIONS, INC. RCA ALASKA COMMUNICATIONS, INC. SATELLITES

General

RCA has an application pending to build three satellites, two of which would be launched, the third to be retained as a spare. Construction is underway on these satellites under several Commission-issued waivers of the requirement for a Construction Permit. Meanwhile, RCA is using

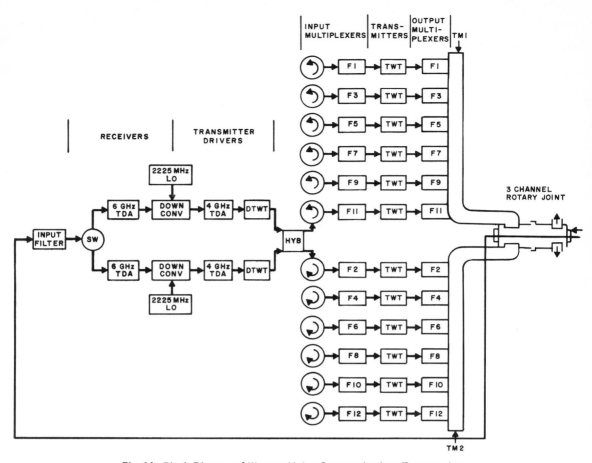

Fig. 14. Block Diagram of Western Union Communications Transponder

TABLE 19. Communications Subsystem Characteristics

Element	Key Performance Features
Overall Subsystem	• 24 36-MHz channels to CONUS, Alaska at 32 dBW • 12 36-MHz channels to Hawaii at 26 dBW
Transponders	• 24 independent channels • lightweight multicoupler diplexers
Antennas	• Cross-polarization isolation Receive 33 dB Transmit 33 dB

transponders supplied originally by Telesat Canada and subsequently by Western Union. Characteristics of the key communications subsystem are shown in Table 19. The basic features of this subsystem which permit the launch of a 24-channel spacecraft satellite by a Delta #3914

vehicle, are the polarization isolation of the overlapping, gridded reflectors, the lightweight filter design, and the high-efficiency TWT amplifiers.

The four-reflector antenna assembly maintains greater than 33 dB isolation between the cross-polarized signals on both the transmit and receive links. This allows re-use of the spectrum so that 24 channels of 40 MHz bandwidth each can be contained in a 500-MHz band.

Characteristics of the other major subsystems are listed in Table 20.

The communications subsystem (see Figure 15) uses the 6/4 GHz common-carrier band, with channel assignments as shown in Figure 16. The plan shows that the frequency spectrum is reused by means of frequency interleaving and orthogonal linear polarization. The 500-MHz band is divided into 12 channels spaced 40 MHz apart, with a second group of 12 interstitial channels cross-polarized with respect to the first group. All 24 channels are available in the Alaska/U.S. beam at full performance (32 dBW EIRP), while the Hawaii beam provides a coupled service at the specified performance level (26 dBW EIRP).

Each of the 24 communications channels is designed to carry single-carrier analog transmissions such as FDM/FM telephony, FM/color TV, or single-carrier digital transmissions including TDMA. Operation is possible also in the multiple-access mode for transmission of multiple-carrier analog or digital signals.

The antenna complement consists of four separate grated reflectors with offset horns mounted on the earth-facing platform. Two

TABLE 20. Spacecraft Support Subsystems Characteristics

Subsystem	Key Performance Features
Command ranging and telemetry	• 248-command capability with false command rate $<10^{-22}$ • 121-channel telemetry capacity with accuracy of $\pm 2.5\%$
Apogee motor	• Mass fraction $\dfrac{\text{Propellant wt}}{\text{total wt}}$ $= 0.939$ • Motor weight to synchronous spacecraft weight ratio = 0.951
Reaction control	• Blowdown monopropellant hydrazine system incorporating surface tension propellant management provides 460 m/s (1500 ft/s) Δ V • Operational redundancy
Attitude control	• Autonomous three-axis body stabilized control system providing pointing accuracies of $\pm 0.16°$ E–W, $\pm 0.21°$ N–S; E–W offset pointing from $0°$ to $\pm 5°$ also provided • Spacecraft is nutationally stable during all modes
Power	• Efficient power system providing over 465 W throughout 10-year design life
Thermal	• Maintains all components in suitable thermal operating environment (e.g., batteries $0°$ to $10°C$ ($32°$ to $50°F$), reaction control $5°$ to $15°C$ ($41°$ to $59°F$) and communications $0°$ to $50°C$ ($32°$ to $122°F$))
Structure	• Lightweight design of 7% of transfer orbit weight • Satisfies all frequency and dynamic envelope constraints of booster

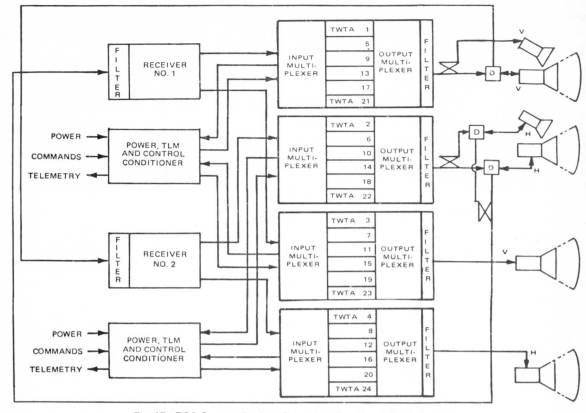

Fig. 15. RCA Communications Subsystem Functional Block Diagram

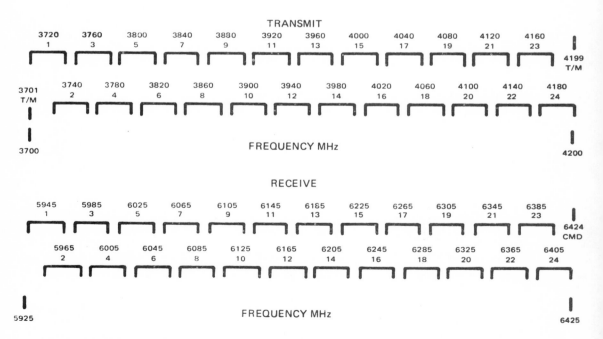

NOTE: NUMBER BELOW CHANNEL CENTER FREQUENCY REFERS TO TRANSPONDER IDENTITY.

Fig. 16. Satellite Frequency Channelization

of the antennas generate vertically polarized beams covering Alaska and the continental U.S.; the remaining two antennas produce identical, horizontally polarized beams, also covering Alaska and the continental U.S. One of each pair operates at 4 GHz only, while the other employs a duplexer for transmit operation at 4 GHz and receive at 6 GHz. In addition, feed horns placed off-axis on the two western antennas produce beams toward Hawaii, and are used in conjunction with three frequency duplexers and couplers to provide the coupled service feature. The resulting interface with the transponder consists of two receive ports and four transmit ports. The two receive ports, one for each polarization, are connected to two separate receivers which amplify the two groups of odd- and even-numbered channels and heterodyne them to the 4-GHz band. At this point, the channels are separated by the input multiplexers and amplified in separate TWT's. The TWT outputs are collected into four groups of six channels by output multiplexers. For both input and output multiplexers, the channels are spaced at 80-MHz center to minimize weight and design complexity relative to a contiguous multiplexer. The resulting four output ports are connected to the four transmit ports of the antenna assembly. Tables 21 and 22 summarize respectively the subsystem EIRP and G/T

allocations, the former based on a 5-W RF output and the latter referred to an equivalent input temperature of 1075 K.

RCA GLOBAL COMMUNICATIONS, INC. RCA ALASKA COMMUNICATIONS, INC. EARTH STATIONS

General

RCA has built, owns and operates earth stations at: Pt. Reyes, California, near San Francisco; Bonsall, California, near Los Angeles; Valley Forge, Pennsylvania; Lena Point, Alaska, near Juneau; Nome, Alaska; Bethel, Alaska; Valdex, Alaska and Putuligayuk ("Put") River, Alaska.

In addition, the Bartlett station at Talkeetna, Alaska, near Anchorage, previously owned by Comsat and used with INTELSAT satellites has been sold and transferred to RCA and will operate in their domestic system.

Functional Characterictics

The earth stations will transmit in the 5925- to 6425-MHz frequency band and will receive in the 3700- to 4200-MHz frequency band.

TABLE 21. EIRP

Parameter	U.S./Alaska (8.4° × 3.2°)				Hawaii (2.6° × 1°)
	Beam Center		Edge of Beam		Edge of Beam
	Direct Channels	Coupled Channels	Direct Channels	Coupled Channels	
TWT Output (dBW)	7.0	7.0	7.0	7.0	7.0
Output Loss (dBW)	1.0	1.0	1.0	1.0	1.0
Coupling Loss (dBW)	0.	0.7	0.	0.7	9.0
Antenna Gain (dBW)	29.9	29.9	26.5	26.5	28.9
EIRP (dBW)	35.9	35.2	32.5	31.8	25.9

TABLE 22. G/T

Parameter	U.S./Alaska (8.4° × 3.2°)				Hawaii (2.6° × 1°)
	Beam Center		Edge of Beam		Edge of Beam
	Direct Channels	Coupled Channels	Direct Channels	Coupled Channels	
Antenna Gain (dB)	29.9	29.9	26.5	26.5	31.5
Coupling Loss (dB)	0.0	0.6	0.0	0.6	10.0
System Temp (dB) (1075 K input)	31.3	31.3	31.3	31.3	31.3
G/T (dB/K)	−1.4	−2.0	−4.8	−5.4	−9.8

+

The stations include high-power transmitters, cryogenically cooled low-noise wideband receivers, uncooled low-noise back-up receivers, ground communications equipment, baseband and multiplex equipment, control, monitoring, test alarms and recording equipment and other equipment and facilities required for multiplex carrier operation. A system block diagram of the earth station is shown in Figure 17.

Initially, transmissions will employ single-channel-per-carrier techniques for up to 150 voice channels, and a wideband FM carrier for a television circuit. Equipment for multi-channel FDM/FM voice circuits can be installed later.

The antenna system consists of a cassegrain antenna with a 10-m (30-foot) main reflector diameter and designed to the minimum G/T objective of 32.4 dB/K.

The radiation pattern of the antennas will be within the envelope described in FCC Standard §25.209. The envelope is defined by $G = 32-25 \log \theta$, where G is the gain in dB and θ is the angle between the axis of the main beam for θ between 1° and 48°. For θ greater than 48°, G is -10 dB.

Technical characteristics of the earth station

TRANSMITTER SYSTEM

Frequence range	5925 to 6425 MHz
Power output/HPA	3 KW, 34.8 dBW maximum/carrier
Frequency tolerance	1 part in 10^6
Bandwidth (tunable)	500 MHz range
Maximum modulating freq.	8 MHz
Frequency deviation	±18 MHz

RECEIVER SYSTEM

Frequency range	3700 to 4200
Bandwidth	500 MHz
Receiving system G/T	32.4 dB/K minimum

ANTENNA SYSTEM

Antenna type	Cassegrain
Effective isotropically radiated power	88 dBW maximum
Minimum elevation angle	10°
Transmit gain at 6 GHz	53.0 dB
Receive gain at 4 GHz	50.8 dB
Radiation in horizontal plane	5.8 dBW/4 kHz (typical)—depends on station location and siting
Beamwidth at 6 GHz	0.36°
Polarization capability	Linear with any orientation—frequency re-use capability
System noise temperature	69.0 K, 18.4 dB maximum

AMERICAN SATELLITE CORPORATION EARTH STATIONS

ASC has built, owns and operates earth stations at Vernon Valley, New Jersey, to serve New York City; Oregon, Illinois, to serve Chicago; Murphy, Texas, to serve Dallas; and Nuevo, California, to serve Los Angeles. These stations can provide links for multi-channel voice, television and/or digital data through transponders supplied by a satellite operator until such time as ASC provides its own satellites.

Characteristics of the earth stations follow. The frequency allocation plan is shown in Figure 18. Transmit and receive functional block diagrams are shown in Figures 19 and 20.

Frequency range:	5925–6425 MHz
Frequency stability:	0.001% (crystal controlled)
Bandwidth per channel:	36 MHz
Power output at antenna feed:	32.5 dBW

Earth Station Communications Antenna, Typical Characteristics

Diameter	11 meters (35 feet)
Gain	50.5 dB at 4 GHz
	52.5 dB at 6 GHz
Beamwidth	0.56° at 4 GHz
	0.36° at 6 GHz
Sidelobe pattern	Meets or exceeds the CCIR recommendation and FCC Standard §25.209 of $32-25 \log \theta$ dB with respect to isotropic for off-beam angles of $1° < \theta < 48°$, and -10 dB with respect to isotropic for $\theta > 48°$
Polarization	Linear: horizontal receive and vertical transmit
Receiving figure of merit	$G/T = 33$ dB/K at 4.0 GHz (minimum)
EIRP	
Transmitting	Main beam: 85 dBW per 36 MHz channel

GENERAL ELECTRIC COMPANY EARTH STATIONS

General Electric (GE) was authorized to build two earth stations: Valley Forge, Pennsylvania, and Pleasanton, California, near San Francisco. Subsequently, GE substituted Daytona Beach, Florida for Pleasanton. These stations are being used in a pilot program for developmental purposes and as part of GE's present leased terrestrial communications network which now provides intra-company communications. The two stations use satellite channels leased from Western Union. Station characteristics are as follows:

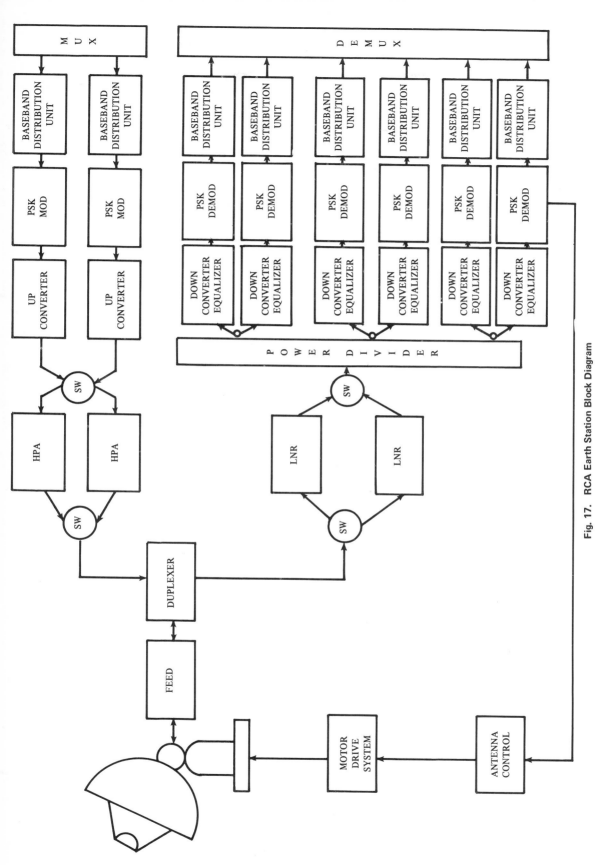

Fig. 17. RCA Earth Station Block Diagram

System Characteristics

Power	3 kW
System G/T	27 dB/K
EIRP per channel	47.8 dBW (uplink)

Antenna Characteristics

Diameter	9 m (26.5 feet)
Polarization	Linear (adjustable)
Gain	52.4 dBi at 6 GHz
	48.5 dBi at 4 GHz
Beamwidth	0.4° at 6 GHz
	0.6° at 4 GHz
Sidelobes	Below CCIR reference pattern

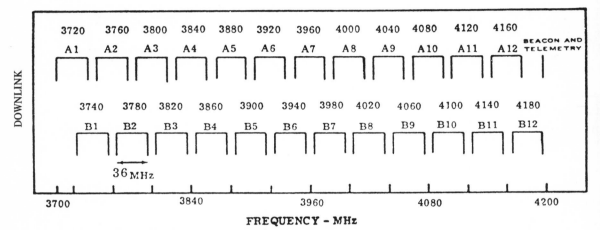

NOTE: NUMBERS BELOW THE FREQUENCIES REFER TO CHANNEL IDENTIFICATION

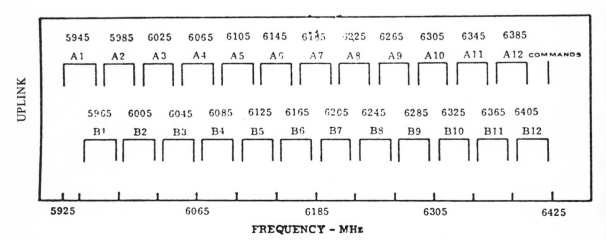

Fig. 18. American Satellite Corporation Satellite Frequency Channelization

52

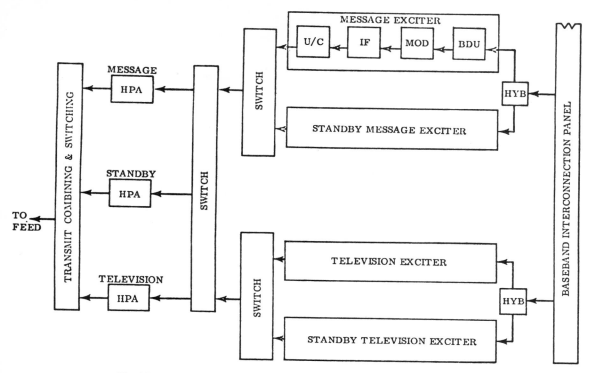

Fig. 19. American Satellite Corporation Earth Station Transmit Function

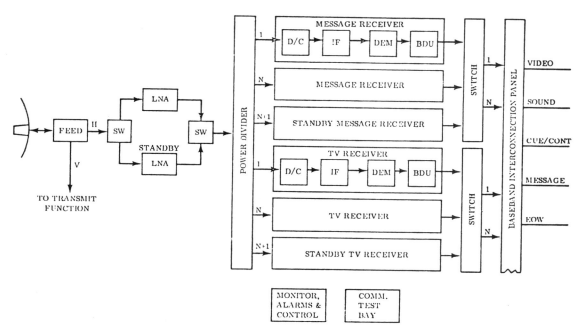

Fig. 20. American Satellite Corporation Earth Station Receive Function

The European Regional Satellite Organization (EUROSAT)

DR. MARCEL R. E. BICHARA
Vibro-Meter AG
Fribourg, Switzerland
(Formerly at CERN, European Organization for Nuclear
Research)

ACKNOWLEDGMENT

The author is greatly indebted to Messrs. Blassel and Lagarde of the Geneva directorate of EUROSAT for their help with this material.

EUROSAT

During the last decade, European organizations demonstrated their ability to master the difficult technological problems inherent in the development of scientific satellites and to manage complex international projects. Today, Europe is undertaking the development of its first applications satellite and, through its participation in the U.S. SPACELAB program, is involved in the field of manned space flight.

The main areas of European participation in space technology are the following:

- Aeronautical satellites to improve communications between aircraft in flight and ground stations, and to facilitate air traffic control on high-density routes. These satellites, originally intended for communications related to aircraft operations, can be adapted to provide communications between the public telephone network and passengers in flight, should the demand for such services arise.

- Maritime satellites: a rapidly developing field up until now characterized by short-term projects such as the MARISAT program (scheduled to become operational in 1975). Longer-term plans are now being prepared under the aegis of the Intergovernmental Maritime Consultative Organization. These plans include consideration of projects such as the European MAROTS program, an adaptation of the INTELSAT V program for maritime use, and a global system operated in the manner of transoceanic submarine cables.

- Meteorological satellites: The first meteorological satellite project is now under way. Termed ESRO METEOSAT it is scheduled for launching at the end of 1976 as part of the Global Atmosphere Research Project (GARP) organized by the World Meteorological Organization.

- Communications satellites: Several projects are being pursued simultaneously. One is the Franco-German experimental satellite SYMPHONIE, the first model of which should be launched early in 1975. (See SYMPHONIE section.) A second is the Italian SIRIO project which should lead to a launch in the beginning of 1975. A third is the OTS (Orbital Test Satellite) undertaken by ESRO in order to verify in orbit the attainment of the requirements defined by the CEPT (European Post and Telecommunications Conference) and by the EBU (European Broadcasting Union). This program should lead to a first launch by the end of 1976. The last is a German television broadcast satellite project.

There are numerous other applications satellite programs of interest to the European community. Some of the missions include: Data collection, observation of earth resources, environmental control and in-orbit manufacturing and processing. Combinations of missions on board the same satellite are being considered.

The development of space systems and of the related technology is a function normally entrusted to special government agencies. In Europe such a function has been performed at a multinational level by two organizations (ESRO and ELDO) which are to be merged into a single European Space Agency (ESA) in the near future. Since neither ESRO, ELDO nor ESA have terms of reference enabling them to carry out commercial operation of systems derived from their development program, a complementary structure had to be set up to provide such a function in a timely manner.

EUROSAT S. A. was founded on January 28, 1972, after more than four years of discussion and consultation among the potential shareholders, who were brought together under the auspices of EUROSPACE. The Company was created to work with the relevant national and international administrations to promote future satellite systems, to manage the space segment of such systems, and to assume the technical, operational and financial responsibility which are thereby implied. The charter of EUROSAT stipulates that the objects of the Company are to:

1. Promote the establishment of operational satellite systems; i.e., for telephone and television, for European regional subscribers by providing all the organizational, managerial, financial and technical services which may be required by authorized clients.

2. Provide authorities for their operation.

3. Promote, outside Europe, the export of the said systems, under the same conditions as in 1.

4. Where necessary, finance the establishment and operation of the said systems, or organize their financing.

EUROSAT now has 98 shareholders of 9 different nationalities. These 98 shareholders comprise 69 industrial firms, 19 banks and financial institutions, and 10 individuals. Expressed in percentage of the share capital (5.2 million Swiss francs), industrial firms hold about 81% of the shares, banks and financial institutions about 18%, and private persons less than 1%.

An approximate breakdown of the shares held within each country is as follows:

France	27%
Germany	30%
Italy	7%
Great Britain	7%
Spain	3%
Sweden	10%
Switzerland	6%
Netherlands	7%
Belgium	3%
	100%

Considerably more financial resources will have to be obtained when EUROSAT manages, on behalf of or in association with public administrations, the space segment of satellite systems. However, the present capitalization of the Company constitutes an excellent basis for rapid acquisition of the additional financial resources required.

Looking towards this future management activity, EUROSAT is preparing itself for close cooperation with the administrations that will use space systems, and with the agencies in charge of technological development and experimental applications programs.

Maritime Satellite Communications

LEO M. KEANE AND DAVID W. LIPKE
COMSAT General Corporation
Washington, D.C.

INTRODUCTION

COMSAT General Corporation, in conjunction with RCA Global Communications, Inc., Western Union International, Inc. and ITT World Communications, Inc., will use two new satellites in geostationary orbit to provide reliable high-quality communications services for both military and civilian maritime users (see Figure 21). Heavy use of the system capacity is expected by the U.S. Navy in the first three years of operation. As Navy service requirements diminish over the lifetime of the satellite, civil maritime use is expected to increase. The following sections describe the facilities and operational aspects of the system.

SHORE STATIONS

A shore station located at Southbury, Connecticut, will be used to interconnect terrestrial facilities with ships in the Atlantic; a similar shore station will be located at Santa Paula, California for Pacific service (see Figure 22). Each shore station has a 14-m (42-foot) diameter antenna which is used to transmit and receive signals to and from the satellite at 6 and 4 GHz. The same antenna is used for transmission (at 1.6 GHz) and reception (at 1.5 GHz) of test and control signals. The shore station figure of merit (G/T) is 31.4 dB/K.

The shore station communications electronics provides the capability for transmission and reception of voice and data signals and pilot tones which are looped through the satellite and used to compensate frequency errors introduced by Doppler shifts and oscillator instabilities. The electronics also provides automatic connection of satellite circuits to terrestrial circuits while monitoring overall system performance.

Initially, each U.S. earth station will be equipped to handle 44 duplex telegraph channels, 22 ship-to-shore simplex channels and six voice

channels. The capacity of the station may be doubled easily by the addition of appropriate modules.

SATELLITE

Two satellites designated "MARISAT" will be used. One will be located at $15°$W and the other at $176.5°$E longitude to provide the coverage shown in Figure 23.

The MARISAT spacecraft design (Figure 24) is based heavily on the flight-proven technology of the INTELSAT IV and ANIK spacecraft. It is a conventional spin-stabilized satellite with a despun antenna array coupled to repeaters on the spinning portion through a three-channel rotary joint. The cylindrical portion of the spacecraft is approximately 2.1 m (85 inches) in diameter and 1.6 m (63 inches) long. The overall length, including all antennas, is approximately 3.8 m (148 inches). Its dry weight is somewhat less than 320 kg (700 pounds). Approximately 300 W of dc prime power will be available at the end of a five-year lifetime. Sufficient battery capability is incorporated into the design to power all subsystems through eclipse. The fixed power available for communications is shared between a UHF repeater used for U.S. Navy service and the 1.6 GHz/4 GHz and the 6 GHz/1.5 GHz repeaters used for civil maritime communications.

A single high-power wideband channel and two narrowband lower-power channels are incorporated in the UHF transponder, the characteristics of which are summarized in Table 23. Separate power amplifiers that can be individually energized are used for each of the wide- and narrow-band UHF channels. To use the total spacecraft power efficiently, a multiple-level TWT has been developed which can operate at three different power levels. If the Navy uses all three UHF channels, the 1.5 GHz TWT operates at its lowest power level; if the

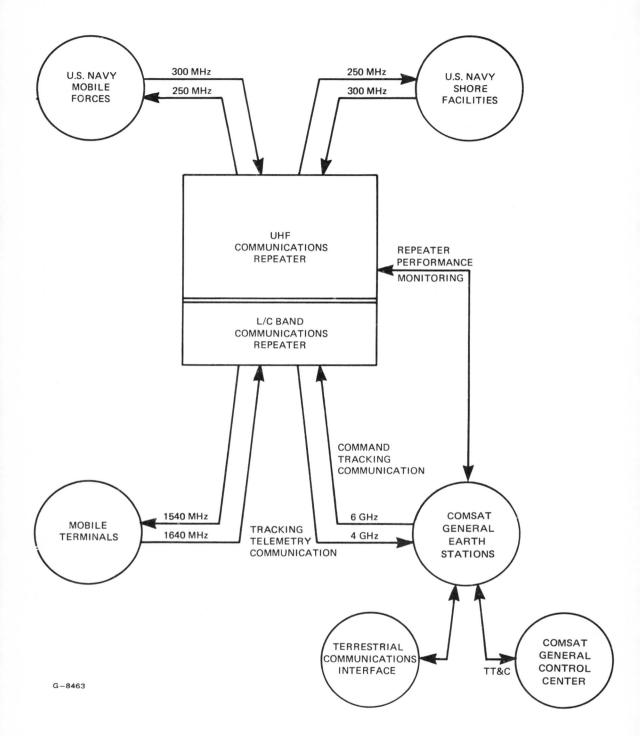

SYSTEM CONFIGURATION

Fig. 21. System Configuration

G-8463

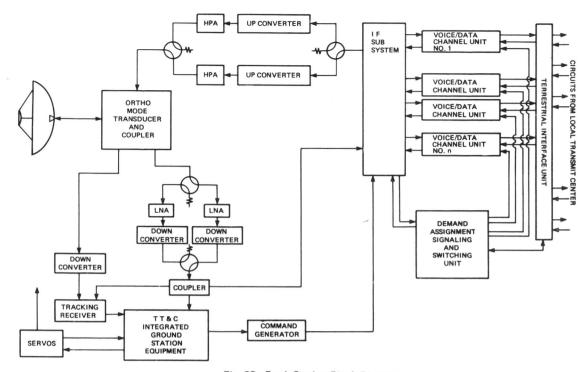

Fig. 22. Earth Station Block Diagram

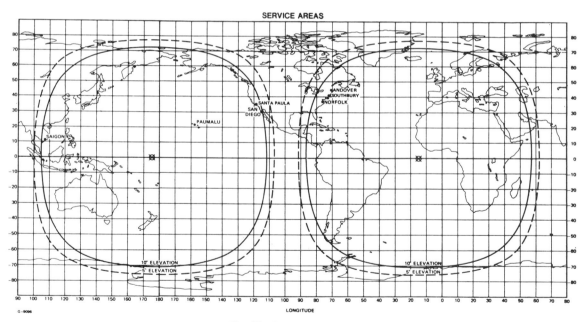

Fig. 23. Coverage Areas

Navy uses only the high-power wideband UHF channel, the 1.5 GHz TWT operates at its mid-power level. When UHF service is no longer required, the TWT can operate at its highest power.

Table 24 demonstrates the simultaneous edge of coverage EIRP's possible for MARISAT. The

1.6 GHz/4 GHz repeater EIRP is independent of the UHF state because the 4 GHz TWT power drain is moderate.

Major characteristics of the civil maritime repeater are shown in Table 25. The 1.6 GHz/4 GHz repeater is a wideband, single-conversion unit

59

Table 23. UHF Transponder Characteristics

Satellite parameter	UHF Service
Receive band	*300 to 312 MHz
G/T	−18 dB/K
Dynamic range for saturation	−151 to −114 dBW/m² for narrowband channels −138 to −114 dBW/m² for wideband channels
Transmit band	*248 to 260 MHz
Saturation EIRP (edge of coverage)	28, 23, 23 dBW in channels 500 kHz, 25 kHz, 25 kHz wide
Normal operation (at saturation)	Single or multiple carrier operation at lessee's option

*One wideband (500 kHz) channel and two narrowband (25 kHz) channels within this band. Translation frequency not necessarily 52 MHz.

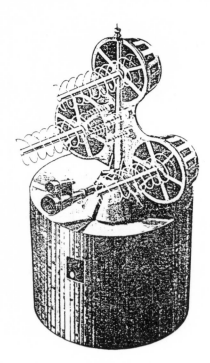

Fig. 24. MARISAT Satellite

Table 24. Simultaneous MARISAT Repeater Operating States

	Saturated Edge of Coverage EIRP (dBW)				
	UHF			1.6-GHz band	4-GHz band
	Wideband	Narrowband 1	Narrowband 2	Shore-Ship	Ship-Shore
All UHF Channels ON	28	23	23	20	18.8
Wideband UHF Channel Only	28	−	−	26	18.8
All UHF OFF	−	−	−	29.5	18.8

Table 25. Maritime Repeater Characteristics

Satellite parameter/service	Shore-to-ship	Ship-to-shore
Receive band	6420 − 6424 MHz	1638.5 − 1642.5 MHz
G/T	−25.4 dB/K	−17 dB/K
Saturation flux density	−87 ± 2 dBW/m²	−96 ± 2 dBW/m²
Transmit band	1537 − 1541 MHz	4195 − 4199 MHz
Saturation EIRP (edge of coverage)	20, 26 and 29.5 dBW	18.8 dBW
Normal Operation	• At saturation • Single or multiple channel per carrier	• Linear • Single channel per carrier

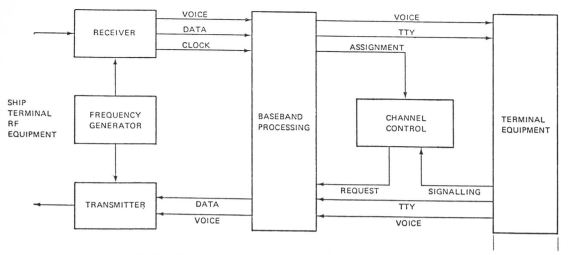

Fig. 25. Ship Terminal Communications Electronics Block Diagram

which normally operates in its linear region. The 1.6-GHz antenna is a near-earth-coverage quad helix located centrally on the despun platform.

The 6-GHz/1.5-GHz repeater is a wideband, dual-conversion unit which operates normally in the hard-limiting region. As the 1.5-GHz TWT has the prime power drain on the spacecraft its efficiency is extremely critical. The 4- and 6-GHz-band antennas are separate horns for transmit and receive and have optimized earth-coverage characteristics.

SHIP TERMINALS

Ship terminal equipment is divided into two parts: (1) an above-deck RF portion consisting of an automatically pointed 1.3-m (4-foot) diameter antenna, a pedestal and a receiver preamplifier, and (2) a below-deck portion consisting of an antenna control unit and the communications electronics needed for voice and data services. The power amplifier is used for both voice and data transmissions. The terminal has a nominal EIRP of 37 dBW and a receiving figure of merit (ratio of antenna gain to system noise temperature) of –4.0 dB/K.

Figure 25 is a simplified block diagram of the ship terminal communications electronics. These units are equipped with automatic features to ensure both operational simplicity and proper network control. One such feature is automatic recognition of messages addressed to the ship with automatic switching to frequencies and/or time slots upon command from the shore station.

CIVIL MARITIME COMMUNICATIONS PLAN

The MARISAT communications system is based on a frequency-division multiple-access (FDMA) plan which permits incremental increases in system capacity as traffic grows and provides the capability for expansion to multiple shore-station operation. Any ship will be able to com-municate directly with any shore station within a satellite coverage area.

High-quality voice transmissions in both ship-to-shore and shore-to-ship directions will employ signal-channel-per-carrier frequency modulation. Because of satellite power limitations, threshold-extension demodulators and compandors are used at the terminals to improve performance. These voice channels, with the compandors removed, can be used for high-quality data transmission (including facsimile) at rates up to 4.8 kb/s. Characteristics of voice transmissions are summarized in Table 26.

The capability for telegraphy transmission is geared to providing real-time, full-duplex telex service over the satellite channels using, as far as possible, standard signalling techniques. Other data rates are accommodated by rate/alphabet conversion using facilities not part of the earth station. In addition to full duplex 50-baud transmission, the chosen signalling techniques permit ship-to-shore simplex operation.

Methods for telegraphy transmission differ for the outbound and inbound links. For the former, twenty-two 50-baud channels are time-division-multiplexed (TDM) and transmitted on a CPSK modulated carrier at a data rate of 1200 b/s. This carrier also contains an assignment channel which informs appropriate ships of incoming calls and automatically causes the ship terminal equipment to switch to the proper time slot (or frequency if voice is to be transmitted). Characteristics of the TDM signals are given in Table 27.

On the ship-to-shore telegraphy link, a simple "open loop" TDMA technique is used. All ship receivers generate timing information from a portion of the TDM signal (TDM signal received by all ships at all times); therefore, no ship-to-ship synchronization is needed. Suitable allowances are made for burst-to-burst transmissions from different ships to account for differences in transmission times resulting from different ship locations. Each burst lasts for about 38 ms and there

Table 26. Voice Transmission

	Shore-to-Ship	Ship-to-Shore
Modulation	FM/SCPC	FM/SCPC
Baseband frequency range (Hz)	300–3000	300–3000
Peak/RMS (dB)	10	10
Syllabic compandor compression ratio	2:1	2:1
C/N_0 (dB-Hz) (edge of beam)	50.4	52.4
Peak deviation (kHz)	12	12
Test tone to noise (dB)	29	31
Equivalent subjective S/N (dB)	28	31

Table 27. Data Transmission

	Shore-to-Ship	Ship-to-Shore	Ship Request
Modulation	2ϕ PSK	2ϕ PSK	2ϕ PSK
Mode	TDM	TDMA (1)	Random
Telex Channels/Carrier	22	22	
Transmission Rate (kb/s)	1.2	4.8	4.8
Bit Error Rate (worst case) (2)	$\leq 10^{-5}$	$\leq 10^{-5}$	$\leq 10^{-5}$
RF Bandwidth (kHz)	1.6	6.25	6.25
C/N Minimum (dB) (3)	10.5	10.5	10.5
C/N_0 Minimum (dBHz) (3)	46.4	48.4	48.4

(1) Open loop synchronization. Burst timing derived from TDM signal.
(2) Nominal bit error rate is considerably less.
(3) Ship at edge of satellite beam.

is an average time of 41 ms between adjacent bursts. The bursts are transmitted at a rate of 4.8 kb/s so that each carrier accommodates 22 simultaneous 50-baud messages.

System access from ship terminals is by means of a random access "request" channel. When a ship wishes to originate a call, a short burst of information is sent to the earth station. A separate rf carrier is shared by all ships for this purpose. Request messages have the same modulation characteristics as TDMA signals and use common ship terminal equipment. Characteristics of

the request and TDMA signals are shown in Table 27.

The capacity of the system for civil maritime use is dependent on the extent of U.S. Navy usage. With full Navy service, the 1.5-GHz radiated power is 20 dBW and the system has the following capacity in the shore-to-ship direction:

● One voice carrier capable of supplying an equivalent subjective average signal-to-average noise ratio of 28 dB

● Two TDM carriers capable of providing 22 telex channels each with a bit error rate $\leq 10^{-5}$.

With higher 1.5-GHz power radiated from the satellite, the system capacity can be increased by the provision of more carriers of the types described. Some or all of the added power could be used also to improve the quality of the existing carriers.

The capacity on the ship-to-shore link should be adequate for 14 equivalent voice channels in all cases, since the available 4-GHz spacecraft power does not change.

SUMMARY

The design of the MARISAT communications system is based on a number of factors which are unique to early satellite capabilities but which will permit the maritime satellite communications service to develop in a logical and orderly manner with eventual expansion into a widely used international service.

Mobile Services

H. L. WERSTIUK
Communications Research Centre
Ottawa, Canada

MIKE A. PROCTOR
Department of National Defence
Ottawa, Canada

THE REQUIREMENT

Communications satellite systems for mobile service have the potential of providing the user with vastly improved telecommunications. High-quality, reliable voice communications via the satellite system, with direct and immediate access to the terrestrial communications network, is a capability far above that provided today through conventional means. In addition to basic telephony, numerous other services are possible. Such services could include data transmission, teletype message services, facsimile, weather and other safety related communications, distress alerting, and position determination services.

The scheduled airline operators, through IATA (1), have established a broad policy which expresses a requirement for a reliable, static-free, long-range communication system providing world-wide voice and data services for air traffic control (ATC) and airline company operations. Position determination services could also be provided if required. Within and near land-mass areas the ATC communication requirement is presently met through a network of conventional Very High Frequency (VHF) radio systems. However, beyond about 200 miles (320 km) from coastlines, full reliance must be placed upon long-range High Frequency (HF) radio communication. The limitations of the HF radio band and its susceptibility to interference are well known. An Aeronautical Satellite (AEROSAT) system has the potential for meeting the stated requirement. However, the special installation requirements for avionics equipment and the economics of airline operations demand a careful examination and proof of both economic and technical feasibility.

A prime operational maritime need is for rapid and reliable communications at all stages of a voyage to ensure efficient and safe operation (2, 3). At present the communications situation for maritime interests is similar to that of the air carriers with VHF and MF providing short-range communications and HF providing long-range communications. Owing to propagation path variability and radio watch schedules, the average delay in transmitting messages to and from ships is nearly six hours according to a study performed by Automated Marine International (AMI) for the U.S. Coast Guard (4). Given the rapid growth of marine radio traffic, the long-term solution having the most potential is the use of a Maritime Satellite (MARISAT) communications system.

DEMONSTRATION OF FEASIBILITY

The general feasibility of satellite communications and position determination has been demonstrated by the Applications Technology Satellite (ATS) program. Detailed evaluation of propagation multipath to aircraft and the performance of different modulation schemes in a multipath environment has been carried out using satellites and balloon-borne satellite simulators. Johnson (5) has reported multipath measurements over an aircraft-satellite link in the frequency range of 225-8000 MHz. Chinnick (6) has reported measurements of the time-spread characteristics of sea-reflected signals on an aircraft-to-satellite link in the 1.6-GHz band. Brown (7), has performed experiments in this same band using a balloon-borne satellite simulator at elevation angles between 5 and 13°.

He observed peak-to-peak multipath signal fading over water of 23 to 46 dB with horizontal polarization, 9 to 28 dB with vertical polarization, and 8 to 15 dB with circular polarization. Encouraging voice intelligibility results were achieved at low signal levels during modulation method comparison tests. A delta-coder with phase modulation appeared to give the best results under these multipath conditions. Wishna (8), during the same experiments, reported aircraft position determination accuracies of a few hundred meters using pulse and CW ranging techniques. Data transmission at 1200 and 600 bits/s also was successful. Sutton et al (9) carried out a series of extensive over-ocean multipath measurements and one-way tone ranging in the 1545-1655-MHz band, using the ATS-5 satellite. Amplitude characteristics, polarization, power spectral density, and selective fading properties were measured for the sea-reflected and composite signals.

The technical feasibility of aeronautical satellite communications has been shown by this and other work. However, the importance of aircraft antenna characteristics at low elevation angles has been highlighted. An antenna system exhibiting good circularity and multipath discrimination at low elevation angles would appear to be essential.

Kaiser and Cacciamani (10) have described a shipboard satellite communications experiment on the Queen Elizabeth II. The trial demonstrated the feasibility of installing a relatively small (2.5-m, i.e., 8-foot) SHF antenna with a simple open-loop pointing system and gyro stabilization. A successful link was established from the ship via the INTELSAT IV satellite to a shore terminal in the U.S.A. The communications capability was demonstrated by transmitting telephone, teletype, computer access signals, facsimile and biomedical data.

Harris and Heaviside (11) have described a small, lightweight, shipboard terminal of limited capacity operating at 7 and 8 GHz. Two stabilized tracking antennas, each 1-m (3-foot) diameter were used. A second version of this system uses a 2-m (6-foot) antenna. The majority of the active equipment is housed in a small equipment cabin designed to be easily and rapidly installed on a prepared deck position.

These and other implementations demonstrate the feasibility of shipborne satellite communications. However the size, performance, installation requirements, and cost of the antenna system are key items in the overall economic feasibility of a MARISAT system.

SYSTEM OPERATING FREQUENCY

The optimum band of frequencies to be used for a mobile satellite communications system has been the subject of several studies and much discussion. Hirst (12) has concluded that the optimum frequency for communicating with aircraft is in the region between 150 and 600 MHz. The AMI study (4) concluded that the feasible frequency range for communicating with shipborne terminals appears to be 160 MHz to 1600 MHz with frequencies in the region of 400 to 500 MHz being optimum.

The 1971 World Administrative Radio Conference for Space Telecommunication (WARC-ST 1971) confirmed that civil aeronautical satellite services might, by agreement, share the VHF band (118-136 MHz) with the established service. In addition, exclusive aeronautical and maritime satellite frequencies between 1535 and 1660 MHz were allocated.

In general, it is desirable to use as high a frequency as possible (consistent with economic and technical feasibility) to exploit the greater bandwidth available and enhance the opportunities for system expansion. The AMI study, on comparing the two allocated bands, reported that a system operating in the 1600-MHz band would be higher in quality, more reliable, and would provide better growth capability than a VHF system. However, the cost of mobile terminal equipment operating in the 1600 MHz band would be higher than that operating at VHF.

The WARC-ST 1971 allocated a band near 1600 MHz bands for the satellite-to-mobile forward channels and mobile-to-satellite return channels. Satellite-to-fixed-station links are permitted either in the 4/6-GHz or 12/14-GHz bands. The choice between these bands will depend on the availability of earth-station facilities and coordination difficulties with existing services. Conservation of satellite power by the use of large, fixed, earth stations will probably make it possible to provide a greater number of return channels than forward channels. These additional channels can be used for mobile access signalling and distress alerting.

IMPLICATION OF OPERATING FREQUENCY ON MOBILE TERMINALS

The installation of antennas on aircraft presents both mechanical and electrical problems. It is desirable that the antenna should be flush-mounted to reduce drag. Hemispherical pattern coverage requirements may require more than one antenna. Consequently, there is considerable merit in utilizing small antenna assemblies capable of being mounted on the outer skin of the aircraft.

It is generally considered that a minimum aircraft antenna gain of 4 dB is required to provide reasonable compromises in acceptable circuit quality, satellite power requirements, and system communications capacity. Conversely, 10 dB would be near the maximum since higher gain would require complex tracking arrangements. MacKeller (13) has estimated the total cost per installation (including maintenance cost) for 4-dB, 7-dB and 10-dB antennas. Although the installed cost of high-gain antennas may be greater, user charges may be reduced due to the increased traffic capacity of the system.

Several groups have produced prototype antenna designs approximating the desirable characteristics. Brian and Mark (14) have reported a circularly polarized disc antenna with a half-power beamwidth of $80°$ and a gain of 5 dB.

Sidford (15) has reported an antenna based upon a slot-dipole element with a peak gain of 5 dB and a slot depth of 1/12 of a wavelength. Work on a linear phased array (16) having a nominal gain of 10 dB also has been reported.

A shipborne antenna system should have high gain to minimize the satellite power requirements for the satellite-to-ship communications link. However, an antenna with a beamwidth smaller than the roll angle of the ship must be provided with some form of stabilization. Kirkby (17) has investigated, by means of a computer simulation, a simple ballasted high moment-of-inertia platform. He concluded the semistabilized antenna platform appeared capable of better than ±5° stability for ships ranging from tankers to trawlers. Such a system would allow ships to use a relatively high-gain antenna while avoiding the cost and complexity of a fully stabilized tracking antenna.

FACTORS AFFECTING SATELLITE DESIGN AND OVERALL SYSTEM COSTS

Several extensive studies have been performed to define the characteristics of the AEROSAT space segment. Most of these studies are based upon using a Delta launch vehicle to minimize costs. Vandenkerckhove (18) has described the results of one series of studies. He concluded that the spacecraft design would be almost equally limited by power, mass and frequency spectrum availability. Of the two types of spacecraft stabilization possible, three-axis stabilization was chosen over spin stabilization because of the potentially greater end-of-life prime power capability (1000 W versus 400 W). Vandenkerckhove chose a satellite antenna with three spot beams arranged to provide the geographic coverage required. Within the basic system parameters chosen, the spacecraft could provide approximately five forward channels. It appeared that almost twice as many return channels as forward channels could be provided.

Martin and Calvit (19) have considered a spin-stabilized satellite (Delta launched) providing five spot beams for a MARISAT system. They investigated the sensitivity of user costs for a range of ship terminal parameters. For a shipborne antenna gain of 18 dB and a noise temperature of 500 K, 100 good-quality voice channels could be provided. They estimated a shipborne terminal with the prescribed antenna gain to cost about $27,000 at a user cost of between $2 to $12 per channel-minute depending on the amount of ship-traffic growth.

EXPERIMENTAL AND OPERATIONAL SYSTEMS

In the near term, extensive experiments for both AEROSAT and MARISAT applications will be conducted on the ATS-6 satellite. Coordinated experiments involving modulation techniques, position determination and mobile station equipment developments are planned.

The completion of negotiations for an international AEROSAT system has met with many delays (20, 21). The current concept includes both the 1600 MHz band and VHF capability for the satellite-to-mobile links, with the first satellite scheduled for launch in 1978. Detailed system planning (22, 23, 24, 25) is proceeding in expectation that the required international arrangements can be concluded in the near future.

Developments have been comparatively rapid in the area of maritime satellite communications. In mid-1973, Comsat General Corporation announced the MARISAT (26) system to provide maritime communications in both the Atlantic and Pacific areas. The system will provide naval communications in the 225-400-MHz band and a civil maritime service in the 1600-MHz band. This service, beginning in early 1975, will initially provide one forward and eight return channels, but this can be increased to eight forward and eight return channels if the military communications capability is phased out.

In mid-1973, the European Space Research Organization began the Maritime Orbital Test Satellite (MAROTS) program (27). The geostationary three-axis satellite is scheduled for launch in 1977, and will be positioned at 12.5°W. One wide-beam antenna will cover most of the Atlantic, eastern Carribean and western Indian Ocean. The MAROTS program is intended to provide a capability for the acquisition of both experimental data and preoperational experience in maritime satellite communications.

The Inter-Governmental Maritime Consultative Organization (IMCO) has held a series of meetings of a Panel of Experts on Maritime Satellites. The objectives of this Panel are to study operational requirements, technical parameters, and financial and institutional arrangements for the establishment of a world-wide MARISAT system. The Panel has considered the establishment of a MARISAT system using a dedicated satellite system approach, and has considered the INTELSAT studies for providing a maritime satellite service using multipurpose INTELSAT V satellites. The dedicated system approach would allow placement of satellites for optimum maritime coverage and allow greater flexibility to accommodate traffic growth. The multipurpose approach, however, would show potential economic advantages, especially for the lower rates of traffic growth and fitting of ships with satellite communications equipment.

References

1. N.G. Anslow and J.O. Clark, "Long Haul Airlines and Satellite Communications," *IEEE Conference on Satellite Systems for Mobile Communications and Surveillance*, 13–15 March 1973, Conference Publication No. 95, p. 116.
2. J.D. Parker, "Maritime Needs for Communication and Navigation via Satellite," ibid, p. 180.
3. G.J. MacDonald, "Operational Requirements for a Maritime Satellite Service," ibid, p. 154.
4. "Automated Maritime International, A Study of Maritime Mobile Satellites," United States Coast

Guard, Report No. DOT-CG-00505A, November 1, 1970.

5. A.L. Johnson, "Measurement of Airborne Propagation Anomalies," *1972 National Telecommunications Conference,* IEEE publication 72 CHO 601-5-NTC.

6. J.H. Chinnick, "Spread-Spectrum Measurements of Sea Reflection Characteristics in an L-Band Satellite-Aircraft Communication Link," *IEEE Canadian Communications and EHV Conference,* Montreal, November 1972.

7. D.L. Brown, "ESRO Aerosat Experiment using Stratospheric Balloons," IEEE Conference ibid.

8. S. Wishna, "Balloon-Aircraft Ranging, Data and Voice Experiment," IEEE Conference ibid.

9. R.W. Sutton, E.H. Schroeder, A.D. Thompson, S.G. Wilson, "Satellite Aircraft Multipath and Ranging Experiment Results at L-Band," *IEEE Trans. Communications,* pp. 639-647, May 1973.

10. J. Kaiser and E.R. Cacciamani, "A Shipboard Satellite Communication Experiment," IEEE Conference ibid.

11. G. Harris and L.W. Heaviside, "Naval Satellite Communications Terminals," IEEE Conference ibid.

12. D. Hirst, "Factors Affecting the Frequency Chosen for Aircraft to Satellite Communications," IEEE Conference ibid.

13. A.G. MacKeller, "Factors Relating to the Choice of Antenna Characteristics for the Aircraft Terminal in an Aeronautical Satellite Communications/Surveillance System," IEEE Conference ibid.

14. D.J. Brian and J.R. Mark, "The Disc Antenna, A possible L-Band Aircraft Antenna," IEEE Conference ibid.

15. M.J. Sidford, "A Radiating Element Giving Circularly Polarized Radiation Over a Large Solid Angle," IEEE Conference ibid.

16. H.L. Werstiuk, J.D. Lambert, L.A. Maynard, J.H. Chinnick, "UHF Linear Phased Arrays for Aeronautical Satellite Communications," AGARD Conference on Antennas for Avionics, Munich, Germany, November 26-30, 1973.

17. R.J. Kirkby, "A Simple Stabilized Antenna Platform for Maritime Satellite Communications," IEEE Conference ibid.

18. J.A. Vandenkerckhove, "Major Factors Influencing the Design of an Aerosat System for Evaluation," IEEE Conference ibid.

19. E.J. Martin and T.O. Calvit, "L-Band Satellite Systems for Mobile Applications" IEEE Conference ibid.

20. H. Falk, "Toward Better Over-Ocean ATC," *IEEE Spectrum,* pp. 64-65, December 1973.

21. D. Israel, "Aerosat Overview," *EASCON '73,* Washington, D.C., September 1973.

22. J. Woodford, "Aerosat Performance Specifications," *Eascon '73* ibid.

23. F.S. Carr, "Aerosat Ground Environment," *Eascon '73* ibid.

24. J. Gutwein, "A First Generation of L-Band Avionics for Aerosat," *Eascon '73* ibid.

25. J.A. Scardina, "Description and Cost of a Satellite-Based Oceanic ATC System," *Eascon '73* ibid.

26. M.C. Nilsen, "MARISAT–The U.S. Maritime Satellite System for Launch in 1974," *24th International Astronautics Conference,* Baku, October 1973.

27. J.A. Vanderkerckhove, J.P. Contzen, "The ESRO MAROTS Programme," *Eascon '73* ibid.

Thin Route Services

P. ROSSITER
TELESAT Canada
Ottawa, Canada

GENERAL

The major role of a thin route service in a satellite communications system is to provide voice communications to small isolated communities in large sparsely populated areas. Northern Canada, for instance, has many such communities.

Typically, these communities
- have populations under 1,000
- have primitive or non-existent telephone communications at present
- require interconnection (a) with major population centers, (b) with similar isolated communities
- can initially support a small number (typically two) of voice circuits
- have a limited future expansion requirement.

Under these conditions, the satellite access techniques chosen will normally be single channel per carrier frequency division multiple access (SCPC/FDMA) since these provide for economical earth-station design together with simple voice circuit add-on expansion capability.

The simplest useful thin-route network structure is one in which all individual remote stations communicate on preassigned frequencies with a master station, the latter being interconnected with the terrestrial telephone network. In this case, a connection between a remote station and the master station is a single satellite hop, and a remote-to-remote connection is double-hop.

Double-hop connections have three basic disadvantages; namely, twice the transmission delay, reduced message quality due to the tandem modulation/demodulation and signal processing and inefficient use of the satellite power and bandwidth resource. However, the double-hop technique is still commercially attractive where circuits are lightly loaded and network traffic patterns are predominantly remote-to-master rather than remote-to-remote.

Double-hop connections could be avoided at the expense of system complexity by the use of demand assignment techniques such as used in the SPADE (1) system, or by preassigning a number of circuits between remote locations.

SIGNAL PROCESSING AND MODULATION TECHNIQUES

Careful evaluation (2, 3) of signal processing and modulation techniques is required in thin-route systems in order to achieve a commercially viable compromise between grade of service, system capacity and earth-station size in a system which for present generation satellites, at least, is power limited.

The major contenders are:
1. Analog
   FM with companding and threshold extension options
2. Digital
   PCM - PSK
   DM - PSK (DM = delta modulation)

In the digital systems DM is preferred over PCM for two reasons:

1. DM exhibits superior audio S/N performance to PCM at moderate (30–40 kb/s) transmission rates,

2. DM degrades more gracefully than PCM at high bit-error rates (BER) and is intelligible even at 10^{-1} BER.

Direct comparison of FM and DM systems on a noise-performance basis is not entirely satisfactory (4) because the nature of the noise is different in each case; i.e., additive noise vs. idle noise and quantization noise. Comparative subjective testing is an important step in the evaluation procedure.

Since both FM and DM-PSK systems may be made economical of satellite power, both techniques will be used, and the final choice will be made on the basis of other properties, such as the following:

1. Resistance to intelligible interference in a high intermodulation-noise environment
2. Residual FM-modulation, local oscillator phase noise
3. Frequency stability
4. Ability to carry classes of traffic other than voice; i.e., voice-band data, high-speed data, signalling and supervision, facsimile, etc.

EARTH STATION DESIGN

The key to the successful exploitation of thin-route systems is simple, economical design for the remote stations.

The capital cost of the earth station is sensitive to antenna and building size, civil works, environmental control design, redundancy philosophy, and communications equipment design.

Operating costs will depend on cost of prime power, and the intervals at which maintenance both scheduled and unscheduled is required. Thin-route terminals are usually unmanned and local technical assistance is not usually available.

TELESAT THIN-ROUTE SYSTEM

Anik

A specific example of an operating Thin-Route (TR) system is the Telesat Canada system (4) which commenced commercial operation in February 1973 via the Canadian domestic satellite (ANIK 1) (5). One RF channel (36 MHz BW) in the 4/6-GHz satellite is dedicated to the Thin-Route Service. Full Canadian coverage is provided by the spacecraft antenna system.

Earth Stations

The Telesat system uses 40 kb/s DM, 20 CPSK FDMA techniques and is now in service in 17 com-

munities in the Canadian far North. The network master station is at Allan Park in southern Ontario. The primary parameters of the remote and master stations are summarized in Table 28.

Initially, each remote site is equipped with two voice circuits with plug-in expansion capability for up to eight voice circuits. Allan Park is equipped to operate on a preassigned frequency basis with the northern stations.

Voice Performance

The voice-circuit quality is jointly determined by the codec design and the system operating BER. The major parameters of the codec at low BER are given in Table 29.

Channel Performance

The 36-MHz transponder bandwidth is subdivided into 600 individual Thin-Route channel frequencies. Nominal channel bandwidth is thus 60 kHz. Channel selection is performed by IF frequency synthesizers and is fixed at present, although provision exists for demand-assigned frequency programming.

Frequency-stability requirements are met by the use of a band-center pilot tone transmitted from Allan Park with spectrum centering performed at all receivers. In addition, all transmitters are equipped with ultra-high-stability frequency references.

The system BER is a function of the received E_b/N_o (energy per bit/noise-power density) and is designed to be less than 10^{-3} for 99.9% of the time.

Implementation margins for the modulator/demodulator are typically from 2 to 2.5 dB when measured in a satellite loop.

Differential encoding is employed.

TABLE 28. Earth Station Primary Parameters

PARAMETER	NORTHERN THIN-ROUTE STATIONS	ALLAN PARK THIN-ROUTE MASTER STATION
Transmit frequency	5985 ± 18 MHz	5985 ± 18 MHz
Receive frequency	3760 ± 18 MHz	3760 ± 18 MHz
Antenna	9-m (26-ft) diameter (non-tracking)	30-m (100-ft) diameter (tracking)
Transmit gain	50.5 dB	62 dB
Receive gain	48.5 dB	59 dB
G/T	22 main, 19 standby	37 dB/K
LNA noise temp.	300 main, 700 standby	100 K
Transmitter	2 × 35 W TWT hybrid combined at output	1.5 kW klystron (redundant multicarrier)
Power - Main	Commercial AC	Commercial AC
Back-up	4 hr lead-acid battery	Diesel/battery
Staffing	Unmanned	Manned - 24 hrs.

TABLE 29. Major Parameters of Thin-Route Delta Codec (low BER)

PARAMETER	PERFORMANCE	
Sampling rate	40	kb/s
Idle circuit noise	28	dBrncO
Dynamic range [for signal-to-quantizing noise (S/Q) $\geqslant$ 25 dB measured at 1 kHz]	$\geqslant 30$	dB
Maximum S/Q (measured at 1 kHz)	$\geqslant 30$	dB
Linearity (gain tracking)	± 1	dB
Frequency response (measured at -10 dBm), referred to 1 kHz		
300 – 3400 Hz	+1, −10	dB
300 – 3000 Hz	+1, − 6	dB
600 – 2400 Hz	+1, − 3	dB

TABLE 30. Data Performance

CLASS OF DATA	DATA INTERFACE	E_b/N_o (40 kb/s PSK)	DATA BER
2400 b/s (voiceband)	CODEC	12	10^{-6}
40 kb/s (high speed)	MODEM	13	10^{-7}

Signalling and Supervision

Signalling and supervision is performed by Telesat's customer, Bell Canada, using single frequency (SF) techniques in the 2600 Hz band. Reliable signalling is accomplished at bit error rates as high as 10^{-3}.

Data Transmission

The Telesat TR system is capable of carrying both low speed data (2400 b/s) and high-speed data (40 kb/s) - the latter being made possible by interfacing directly with the PSK modem and the former by interfacing with the delta codec.

Typical measured data performance is given in Table 30.

Shelter Design and Civil Works

In the northern stations, the electronics equipment is housed in a 6 × 2.5 × 2.5 m (20 × 8 × 8 ft) fiberglass shelter. Outside ambient temperatures may range from $-46°$C to $27°$C ($-50°$F to $+80°$F) over the year so that active thermal control is necessary to maintain the inside ambient temperatures at $21°$C $\pm 11°$C ($70°$F $\pm 20°$F.)

A thermostatically controlled heater/cooler unit is provided, together with intake fans, to accomplish this climate control.

The 8-m (26-ft) antenna is mounted on poured concrete foundations alongside the shelter and is connected to the communications equipment via a waveguide run.

Manual de-icing is provided for the antenna feed horn window and sub-reflector by means of infrared lamps. No de-icing provisions are made for the main reflector.

FUTURE GROWTH OF THIN-ROUTE SYSTEMS

The prospects for development and expansion of Thin-Route systems depend largely on the ability to produce cost-effective earth stations.

Future growth may bring the following:

1. Small antennas of 4.5-m (15-ft) diameter which must provide both adequate gain and improved off-axis discrimination against adjacent interfering satellite systems

2. A critical examination of site (survival and pointing accuracy) requirements

3. Clearly defined and limited expansion capabilities

4. Use of voice activation techniques

5. Use of all-transistor low-noise RF amplifiers

6. Critical examination of com nications equipment redundancy requirements.

Improved designs together with increasing user requirements will guarantee the future of Thin-Route services both in Canada and other countries of the world.

References

1. B.I. Edelson and A.M. Werth, "SPADE System Progress and Application," *Comsat Technical Review*, vol. 1, Spring 1972.

2. P.M. Norman and D.E. Weese, "Thin Route Satellite Communications for Northern Canada," *ICC '71 Conference Record.*

3. P.M. Boudreau and N.G. Davies, "Modulation and Speech Processing for Single Channel per Carrier Satellite Communications," ibid.

4. P. Rossiter, "System Aspects of the Initial Telesat Thin Route Satellite Communications System," *ICC '73 Conference Record.*

5. R.M. Lester, "Canadian Domestic Satellite System Applications," presented at *1973 IEEE Intercon.*

Broadcast Satellite Service

JAI P. SINGH
Indian Space Research Organization
Bangalore, India

During the past few years interest has grown in the use of satellites for direct broadcast transmission to earth. (1–8)

The Broadcasting-Satellite Service (BSS) is a space service in which signals transmitted by space stations (satellites) are intended for direct reception by the general public. Broadcasting-satellite services or systems can be divided into two categories: systems that allow individual reception; and systems that are designed for community reception. In the case of individual reception or direct-to-home type broadcasting satellite systems, the emissions from the satellite are sufficiently powerful for reception through simple domestic installations. Community-reception type BSS systems use receiving equipment which may be installations having large antennas and intended for group viewing and listening or for local distribution of signals by cable. In some cases rebroadcasting to limited areas is involved.

Two reasons for using satellites for broadcasting are:

1. Fast introduction of service
2. Low cost of reaching a large number of receivers.

In the late 1960's there was great interest in voice-broadcast satellite systems. Two major systems studies were devoted to the investigation of technological and cost factors associated with unmanned satellites for direct broadcast of aural material to unmodified home receivers in the frequency bands 15–26 MHz, 88–108 MHz, and 470–890 MHz (9–12). The studies concluded that the voice broadcast missions were technically feasible for a launch in the early 1970's, provided effort on items needing long development and testing times was started in the immediate future (10). However, it was found that high-frequency broadcasting satellites were not economically competitive with terrestrial systems.

On the other hand, VHF voice broadcasting satellite systems were found to be cost competitive with terrestrial systems and it was suggested that the best use of such satellites would probably be in broadcasting to the increasing number of low-cost, small, and inexpensive portable AM-FM receivers. However, satellites for VHF transmissions were also discovered to be large and complex due to the frequency band used. The 1971 World Administrative Radio Conference for Space Telecommunicating (WARC-ST) did not accept any proposals for aural broadcasting from space in either HF or VHF bands and thus forestalled options for direct voice broadcasting to augmented as well as unaugmented receivers in these frequency bands. However, the U.S. education authorities and especially National Public Radio (NPR) continue to be interested in community reception voice broadcasting from space and it is quite conceivable that such a service may be provided in the future in the 2.5- or 12-GHz BSS bands for local dissemination through cable or local broadcast stations.

Interest in direct television broadcasting from space dates to the early 1960's. In the middle and the late 1960's detailed and comprehensive studies were made to investigate technical and cost factors associated with the implementation of direct television broadcast systems (13–19). These and subsequent studies (16,17,20,21,22) treat operating frequencies, television standards and signal formats, sound-channel placement, interference with terrestrial as well as other satellites sharing the same frequency bands, modulation techniques, receiving terminal cost and technology, satellite technology (antenna, power and attitude control and station-keeping subsystem), and launch vehicles.

Early proposals for direct television broadcasting from space were designed around unaugmented conventional receivers in the Ultra High Frequency

(UHF) band. These studies indicated that systems would not be operational in the 1970's because the required transmitter power at the satellite (22-24) could not be produced. However, the decision of the 1971 WARC-ST to allow only FM by the Broadcasting-Satellite Service in the frequency band 620–790 MHz precluded the possibility of broadcasting to conventional home receivers without some augmentation such as a modulation converter.

Most of the current interest in direct television broadcasting is centered around community reception systems for educational purposes which might allow the introduction of a multi-channel system by the mid 1970's.

Most of the future broadcasting satellites are expected to be placed in geostationary orbit to provide a continuous service to fixed areas on earth as well as to minimize receiving terminal cost (23,24). If far-northern coverage is desired (such as the Soviet Union), inclined highly elliptical orbits become suitable.

As a consequence of the 1971 WARC-ST allocations, current operating frequencies are limited to the 620–790 MHz, 2500–2690 MHz and 11.7–12.2 GHz bands. The first two bands are limited in radiated power. Available downlink power is the limiting constraint on system performance for area coverage from satellites to small terminals. Minimum cost of the earth terminal, as determined by its antenna, indicates usage of the lower end of the 1–10 GHz band (25). However, where the lower frequencies are heavily used by the terrestrial services, as in the case of the UHF-TV band in the U.S., the reception of a satellite FM broadcast signal becomes difficult in the presence of strong local TV or mobile transmissions without the use of adaptive or adapted arrays.

Table 31 presents the details of three representative systems, at 800 MHz, at 2.5 GHz and at 12 GHz (23). All systems provide television at 525 lines/60 frames per second with 4.2-MHz video bandwidth, TASO grade 1 picture quality, frequency modulation, a combined noise weighting and preemphasis improvement of 12.5 dB and a carrier-to-noise ratio of 13 dB. The earth terminals for 800 MHz and 2.5 GHz are assumed to have antennas of 3-m (9.5-foot) diameter and system noise temperature 1100 K. At 12 GHz the system noise temperature is assumed to be 1500 K and antenna diameter as 1.2 m (4 feet). The receiving antenna diameter is set because a narrower beam is not compatible with the goal of providing a low-cost non-tracking receiving terminal, with the current satellite technology.

The constraint on antenna diameter, as well as increased atmospheric attentuation, places the 12-GHz system at a considerable disadvantage as indicated by the effective radiated power requirements of the three systems. Also the cost of a 12-GHz receiver would be somewhat higher than that of 800-MHz and 2.5-GHz receivers due to the high cost of the preamplifier and mixer.

Operation at 12 GHz does, however, offer certain advantages including a great deal of flexibility in siting receiving terminals. The absence of a power flux-density limit in this frequency band can result in lower total system costs, since high-power satellites can be used with many low-cost receiving stations. Few terrestrial users are in the band and thus interference considerations do not restrict system design. Furthermore, the bandwidth to accommodate about 12 TV channels is available at 12 GHz, and therefore it is possible to use satellite-borne multiple narrow-beam, shaped-pattern antennas that could operate at reduced power (by eliminating the spillover in unwanted areas) and would facilitate reuse of the spectrum (23,26).

The ability to design and implement a broadcasting satellite system is closely related to current and projected satellite technology in the areas of prime power, transmitters, thermal control, antennas and launch vehicles. The requirements imposed on satellite technology by the community-reception systems for the mid-1970's are such that they can be met by current hardware development programs of NASA, such as the 10-m (33-foot) ATS-6 space-deployable antenna and associated 80-W UHF transponder for the India-NASA experiment (20,27); sun-oriented solar cell arrays providing 1.5 kW of prime power and extendable to 10–15 kw; transmitters employing TWT's generating 200 W at 2.5 GHz with efficiencies greater than 50 percent; and super-efficient 1–4 kW klystrons and TWT's for operation at 12 GHz (28). Launch vehicles are technically feasible now for either community-type systems or direct-to-home broadcasting to augmented receivers; however, the cost per launch for direct-to-home broadcasting is high at present. Reusable space shuttles may make the latter more economical.

There is general agreement now that economics rather than technology is the limiting factor in the design of either AM or FM television receivers for direct reception. NASA has supported and is continuing to support receiver development studies to determine the characteristics and costs of the FM receiving systems and front-ends (22,28). Development work on 12-GHz AM receivers has taken place particularly in Europe (16). A novel, high-performance (noise figure less than 4.5 dB) 12-GHz wideband receiver capable of handling 3-4 TV channels has been developed at the NHK Research Laboratory in Japan (33). This design seems to have the potential for making the receiver system cost as low as $150 per unit when manufactured at a level of 100,000 units (34). Table 32 presents manufacturing cost estimates for receiving units (28). Also shown in the table are the typical ranges of estimated satellite transmitter power per TV channel. These power levels are determined by desired coverage area, receiving antenna size and desired picture quality.

NASA's sixth Application Technology Satellite, (ATS-6) launched on May 30, 1974, has flight-tested an 80-W, 860-MHz transmitter which will be used for community TV experiments in India (7,26,27,36). Also on board ATS-6 is a pair of 2.5-GHz 15-W transmitters for satellite ETV trans-

TABLE 31. Representative System Parameters (23)

Parameters	800-MHz System	2.5-GHz System	12-GHz System
Frequency deviation	9 MHz	9 MHz	9 MHz
IF bandwidth	26.5 MHz	26.5 MHz	26.5 MHz
Carrier power	−107.6 dBW	−106.9 dBW	−106.3 dBW
Effective antenna area	3.9 m^2	3.9 m^2	0.6 m^2
Flux density (at receiver)	−113.5 dBW/m^2	−112.8 dBW/m^2	−104.1 dBW/m^2
Field strength (at receiver)	32.5 dB	33.2 dB	41.9 dB
Rain loss	0 dB	0 dB	3 dB
EIRP (at −3 dB pt.)	53 dBW	53.7 dBW	65.4 dBW
Satellite antenna beamwidth	3°	3°	3°
Satellite antenna diameter	8.4 m (27 feet)	2.8 m (9 feet)	0.6 m (1.9 feet)
Transmitter line loss	1 dB	1 dB	1 dB
Transmitter power	80 W	94 W	1.4 kW
Prime power	146 W	172 W	2.5 kW

TABLE 32. Estimated Receiver Manufacturing Costs and Satellite Transmitter Requirements (28)

	Receiver Cost as a Function of Quantity Manufactured		Estimated Satellite Transmitter Power Range (Watts)
	1000 units	100,000 units	
800 MHz–FM, mixer	$40	$25	100–400
2.5 GHz–FM, mixer	45	25	100–400
12 GHz–FM, mixer	85	40	500–2200
Added tunnel diode preamplifier	75	40	About half of power shown above
Receiving antenna	20 D^2	10 D^2	

where D (diameter) is in meters

missions directly to public broadcast stations for rebroadcast, to cable-television system head-ends in remote areas in Montana and Wyoming, and to rural schools in the Rocky Mountain region and Alaska (26). In combination with the 10-m (33-foot) diameter deployable antennas, these transmitters provide an effective radiated power high enough to permit use of low-cost receiving terminals.

The Communications Technology Satellite (CTS), discussed in a previous Section, is scheduled for launch late in 1975 and will also experiment with TV transmission in the 12-GHz band for community-type reception. Its proposed list of experiments also includes broadcast of aural signals to a selected region.

Demonstrations and experiments with ATS-6 and CTS satellites are bound to stimulate greater interest in the development and deployment of domestic as well as regional high-power satellite systems, and in particular, those for group viewing and limited local redistribution. Though currently there are no plans in the U.S. and Canada for such an operational system, it is not unlikely that in the late 1970's high-power satellite systems capable of feeding a large number of TV channels directly to low-cost broadband terminals will be deployed.

India has plans to follow up the ATS-6 community-TV experiment with an operational satellite system to be known as INSAT (30,37). Brazil is also planning a satellite system of its own that will provide for direct reception of satellite signals by school roof-top installations (31). France has prepared an elaborate plan for satellite TV distribution and broadcasting in the French-speaking African countries (32), and there seems to be considerable interest in a 12-GHz satellite broadcast system for western European countries. At least some of these would see actual deployment in the 70's or early 80's.

The Japanese Space Agency, NASDA, has moved ahead in the Direct Broadcast Satellite area through placement of an order in early 1974 for the development and fabrication of a 12-GHz high-power experimental broadcast satellite which is scheduled for launch in late 1977 with the Delta 2914 vehicle. The mission objectives of this project are to provide picture- and voice-transmission tests of a broadcast satellite system operating in the 12/14-GHz band, evaluate performance of ground transmitter/receiver systems and establish satellite control techniques (35). Figure 26 shows the various elements of the project. The satellite under development is to have three-axis stabilization with a pointing accuracy of about 0.1° and a three-year design life with 0.5 probability of success. The onboard power is about 628 W, allowing two simultaneous independent TV channels with 55 dBW EIRP each at 3 dB beam edges with 100 W TWT's and a narrow-beam (1.4° × 2°) spacecraft antenna. The plans call for use of simple ground terminals with 1.6-m (5-foot) antenna with a figure of merit (G/T) of 15 dB/K on the mainland falling under the 55 dBW EIRP contour, and terminals with 24.8 dB/K and 4.5-m (15-foot) antennas in distant areas like Okinawa which fall under the 46 dBW EIRP contour.

While the Direct Broadcast Satellite (DBS) technology has made great progress, as indicated by the emergence of a project like Japan's experimental high-power broadcast satellite, the future of operational DBS is still obscured by the political and legal questions which the United Nations has not yet been able to resolve (38). The issues to be resolved in establishing principles and regulations governing direct broadcasting by satellites involve participation of all interested nations in DBS, signal spillover to the territory of other countries, prior consent of the country concerned before undertaking direct satellite broadcasting to that country, program content, legality of broadcasts, duty and right of consultation with nations which may suffer harmful interference due to the DBS, peaceful settlement of the disputes and copyright and protection of television signals (39).

References

1. "Useful Applications of Earth-Oriented Satellites, Broadcasting," prepared by Panel 10 of the Summer Study on Space Application, National Academy of Sciences, Washington, D.C., 1969.
2. "Report of the Second Session of the Working Group on Direct Broadcast Satellites," Committee on the Peaceful Uses of Outer Space, United Nations General Assembly, Document No. A/AC. 105-66, August 12, 1969.
3. "Broadcasting from Satellites," Working Paper submitted by Canada and Sweden to the Working Group in Direct Broadcast Satellites, Committee on the

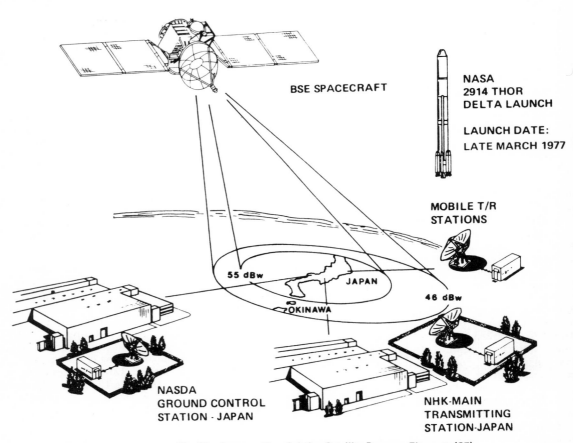

Fig. 26. Japanese Broadcasting Satellite Program Elements (35)

Peaceful Uses of Outer Space, United Nations General Assembly Document No. A/AC. 105-49, February 13, 1969.

4. Gould, R.G., "TV Broadcast from an Earth Satellite," *IRE Transactions on Communications Systems*, pp. 193–201, June 1961.

5. J.L. Hult, "Broadcast Opportunities with Satellites and CATV and their Control in the Public Interest," The Rand Corporation, Santa Monica, California, March 1970.

6. J.L. Hult, "Satellites and Future Communications Including Broadcast," Proc. 13th Annual Meeting of the American Astronautical Society, Dallas, Texas, May 1967.

7. B.S. Rao and R.P. Froom, "Broadcasting from satellites: A powerful potential aid to the new or developing countries," *Telecommunications Journal*, vol. 38–VII, pp. 529–537, July 1971.

8. J. Hanessian, Jr., and J.B. Margolin, "Broadcast satellites: Their potential use for educational purposes and their relationship to international understanding and cooperation," Occasional Paper No. 3, Program of Policy Studies in Science, and Technology, The George Washington University, Washington, D.C., July 1969.

9. R.W. Hasselbacher, "An Evaluation of Voice Broadcast Satellite Systems," *AIAA Second Communications Satellite Systems Conference*, San Francisco, California, Paper No. 68-423, April 1968.

10. P.W. Kuhns, "Feasibility Studies of Direct Voice Broadcast Satellites," Lewis Research Center, NASA Technical Memorandum TM X-1747, Cleveland, Ohio, February 1969.

11. "Voice Broadcast Mission Study," General Electric Company, Philadelphia, Pennsylvania, Rep. 67SD4330, Space Systems Organization, July 1967.

12. "Voice Broadcast Mission Study," Radio Corporation of America, Astro-Electronics Division, New Jersey, Rep. AED-R-3187, May 1967.

13. R.W. Hasselbacher, "An Evaluation of Television Broadcast Satellite Systems," *Journal of Spacecraft and Rockets*, vol. 6. no. 10, October 1969.

14. J. Jansen, et al., "Television Broadcast Satellite Study," TRW Systems Group, Report No. NASA CR-72510, TRW No. 08848-6002-R0-00, Redondo Beach, California, 1969.

15. R.W. Hasselbacher, "Television Broadcast Satellite (TVBS) Study–vols. I–II," General Electric Company, Space Systems Organization, Report No. NASA CR-72578, Philadelphia, Pennsylvania, August 1969.

16. K.G. Freeman, et al., "Some Aspects of Direct Television Reception from Satellites," *Proceedings of the IEEE*, vol. 117, no. 3, pp. 515–520, March 1970.

17. A.K. Jefferis, "Satellite Television Distribution Service from Geostationary Satellites in Multiple-Coverage areas," *Proceedings of the IEEE*, vol. 116, no. 9, pp. 1501–1504, September 1969.

18. R.P. Haviland, "Choices in Space Broadcasting," *IEEE Transactions on Broadcasting*, vol. BC-13, no. 3, pp. 80–86, July 1967.

19. "Technical and Cost Factors that Affect Television Reception from a Synchronous Satellite," Janksy and Bailey Systems Engineering Department, Atlantic Research, Report TR-PL-9037, Washington, D.C., June 1966.

20. P.J. Heffernan, "Television Relay Using Small Terminals (TRUST)," Goddard Space Flight Center, Report X-733-68-163, Greenbelt, Maryland, May 1968.

21. "Television Broadcast Satellite Study," vol. I-4, General Dynamics, Convair Aerospace Division, Report No. GDC-DCF70-002, San Diego, California, December 1970.

22. "Satellite Broadcasting Systems Study," Computer Sciences Corporation, Report No. 4124-011, Falls Church, Virginia, December 1971.

23. A.M.G. Andrus, "Television Broadcasting Satellite Possibilities," *Proc. Mexico International Conference on Systems, Networks, and Computers*, Oaxtepec, Mexico, January 1971.

24. "Broadcasting Satellite Service: Technical Considerations," Working Paper submitted to the Working Group on Direct Broadcast Satellites by the United States Delegation, Committee on the Peaceful Uses of the Outer Space, Document, A/AC.105/50, United Nations General Assembly, New York, February 1969.

25. J.L. Hult, "Sharing the UHF Between Space and Terrestrial Services," The Rand Corporation, Report P-4436, Santa Monica, California, September 1970.

26. L. Jaffe and D. Silverman, "Satellites for Television Distribution," Office of Space Science and Applications, National Aeronautics and Space Administration, Washington, D.C., 1971.

27. J.P. Corrigan, "The Next Steps in Satellite Communications," *Astronautics and Aeronautics*, pp. 36–41, September 1971

28. P.W. Kuhns, "Directions and Implications of Communications Technology," NASA Technical Memorandum TM-X-52911, Lewis Research Center, Cleveland, Ohio, November 1970.

29. N.E. Feldman and C.M. Kelly, "The Communication Satellite–A Perspective for the 1970's," *Astronautics and Aeronautics*, pp. 22–29, September 1971.

30. V. Sarabhai et al., "INSAT–A National Satellite for Television and Telecommunication," *National Conference on Electronics*, Bombay, March 1970.

31. "Project SACI Report No. 111," "Technical Report LAFE-104," Sao Jose dos Campos, Sao Paulo, Brazil, February 1970.

32. "SOCRATE–Television Educative en Afrique," Centre National Etudes Spatiales, Paris September 1971.

33. Y. Konishi and K. Uenakade, et al., "Proposed SHF FM Receiver for Satellite Broadcasting," *Proc. 1973 International Conference on Communications*, vol. II, Seattle, Washington, June 1973, pp. 36–15–36–19.

34. Report of the United Nations Panel Meeting on Satellite Broadcasting Systems for Education (Tokyo, March 1974), pp. 12–13, United Nations General Assembly Document A/AC.105/128, United Nations, April 5, 1974.

35. "A High Powered Experimental Broadcast Satellite Using Tried and Proven Technology," General Electric Space Division, Valley Forge Space Center, Philadelphia, 1974.

36. B.J. Trudell, "Applications Technology Satellite ATS-F Spacecraft Reference Manual," Document X-460-74-154, Goddard Space Flight Center, National Aeronautics and Space Administration (NASA), Greenbelt, Maryland, May 1974.

37. U.R. Rao, "Educational Television in India," *Advances in the Astronautical Sciences*, American Astronautical Society, vol. 30, pp. 97–116, California, 1974.

38. "Report of the Committee on the Peaceful Uses of the Outer Space" Twentyninth Session of the General Assembly, Supplement No. 20 (A/9620), United Nations, New York, 1974.

39. "Report of the Working Group on Direct Broadcast Satellites On the work of its Fifth Session," document no. A/AC.105/127, General Assembly, United Nations, New York, April 2, 1974.

Part 2

Techniques For Expanding Communication Satellite System Capabilities

Modulation and Multiple Access Techniques
Frequencies above 10 GHz
Techniques for Frequency Reuse

Modulation and Multiple Access Techniques

Extracted From
"A Review of Satellite
Systems Technology" (1972)
[IEEE 72CHO749-2 AES]

MODULATION METHODS

General Discussion

The efficiency of a satellite system depends upon the modulation technique used, which consists basically of two parts:

1. A method of assembling the individual channels before they are modulated onto the main RF carrier; i.e., frequency-division multiplexing (FDM) or time-division multiplexing (TDM).

2. A method of modulating the multiplexed baseband onto the RF carrier; i.e., frequency modulation (FM), phase modulation (PM) or phase-shift keying (PSK), etc.

Individual techniques may be used in combination to suit a particular situation and this gives rise to composite techniques such as FDM/FM, TDM/FM or TDM/PSK.

In a multiple-access environment with many earth stations accessing the same satellite transponder, the modulation techniques fall into four areas:

1. frequency-division multiple-access (FDMA)
2. time-division multiple-access (TDMA)
3. spread-spectrum multiple-access (SSMA)
4. pulse-address multiple-access (PAMA)

Each of these is described in the following:

1. FDMA

The available transponder bandwidth in FDMA is divided into a number of non-overlapping frequency bands with bandwidths dependent upon the traffic transmitted on each carrier. The traffic is assembled in FDM form and the composite baseband is used to frequency-modulate the RF carrier.

This technique has been employed extensively in commercial communications satellite systems where the satellites have been power limited. Its efficiency lies in spreading the power over the available bandwidth. Furthermore, FDMA is easy to incorporate because many terrestrial links use FDM/FM.

For multiple access, however, there are several disadvantages. The satellite repeater is a nonlinear device and creates signal impairments such as intermodulation and crosstalk. To reduce these effects the output amplifier has to be "backed-off" to operate in a linear region, and therefore, the available EIRP of the satellite is reduced. Also guard bands between channels are necessary and careful spacing of the carriers (to allow for varying capacities) is required to reduce intermodulation. These effects produce a reduction in satellite transponder capacity. For multiple-carrier operation, the satellite power must be shared among all carriers, and this requires close control of the uplink transmitter power at the earth station. Perhaps the major disadvantage of the technique is that it requires carriers to be allocated permanently to the various earth stations and thereby makes the system difficult to modify with changes in traffic distributions.

2. TDMA

In TDMA the transponder is used exclusively by a single specified carrier during a specified time slot. At any one time, therefore, only one carrier is using the transponder output amplifier (which may be operated at saturation) and earth station power coordination is not required. TDMA is ideally suited for digital-modulation techniques, and systems using this technique have already been proposed (1) and tested.

While this technique is bandwidth-efficient and energy-efficient, it requires the transmission of a clock signal to synchronize the various earth stations' information bursts. This is accomplished without difficulty by the use of realtime computing systems employing integrated-circuit technology.

3. SSMA

The carriers from the earth stations using SSMA are frequency- or phase-modulated in such a way that their spectra occupy the whole of a common bandwidth. One method of generating a spread-spectrum signal is to multiply the sinewave output of a crystal-controlled oscillator by a pseudorandom binary signal which has a basic frequency of half the desired spread-spectrum bandwidth. The pseudorandom binary signal represents the code by which the carrier is "addressed" to the signal for demodulation of the carrier. The pseudorandom generators, therefore, perform the function of addressing as well as spectrum spreading, and for this reason synchronization between stations is required. The message modulation appears on the transmitted carrier as an additional frequency or phase variation.

The advantage of this system is that the spectrum of each carrier approaches Gaussian, which reduces the intermodulation and increases the immunity of the signals to interference. The addressing technique makes for a good flexibility without timing problems. This technique is not normally used because it requires synchronization and complex terminal equipment, transmitter power needs to be controlled accurately, and it gives poor transponder bandwidth utilization.

4. PAMA

PAMA is a method in which time-frequency pulse patterns are generated and used as addresses. The frequency spectrum is split into bands, and pulses are transmitted in the various bands at preassigned times. Reception of signals is accomplished by detection of both frequency- and time-slot positions and addressed receivers can lock on to the transmitted signals in a short time. System synchronization is therefore not required. Matched-filter or correlation detection is used for reception, but PAMA has low energy efficiency. Like SSMA, a large number of codes can be used, which makes PAMA suitable for use in random-access systems. With the advent of the present very high power satellites utilizing power concentration in the form of spot beams, this technique may be useful as an alternative way of increasing transponder capacity.

FDMA and TDMA are well suited for a high-capacity system with few traffic fluctuations, whereas SSMA and PAMA are better suited to the case of a large number of users having small but quickly changing traffic requirements where random access is desirable. For demand-assigned systems a much more flexible version of FDMA; e.g., SPADE (2) or COSIC (3), can be produced utilizing single-channel-per-carrier techniques.

The method of channel assignment has no relation to the multiple-access technique used. A particular type of multiple-access technique may be better suited than another to a particular form of channel assignment, but it is still possible to use the same technique with different forms of assignment. It has been found that where low-traffic routes are concerned, demand assignment gives

vastly improved efficiency, whereas preassignment gives adequate efficiency for heavy-duty routes.

Comparison of Traffic Capacities

There are many modulation/multiple-access techniques possible; to date FM/FDMA has been used in INTELSAT satellites on preassigned circuits and PCM/PSK/FDMA (SPADE) has recently been used as a demand-assigned service. The fully variable PCM/PSK/TDMA system is still under development but this will probably form the nucleus of future satellite systems due to its improved satellite channel capacity (4) and the current trend toward digital systems. In addition to these, delta modulation has been used in the TELESAT system thin-route service. Companded frequency modulation is under consideration both as an alternative to digital modulation (6) and for small-station or thin-route systems (3, 7) where improved satellite utilization can effect simplification of the earth-station equipment. Such a system would employ syllabic compandors to give an improvement in channel quality and satellite capacity.

There have been many calculations to compare transponder capacitors (3, 4, 6), all of which depend critically on the assumptions made. Figure 1 presents one such comparison for various single-channel-per-carrier methods which are typical of both preassignment and demand-assigned systems. The conclusions from this comparison are that PCM/PSK/TDMA is preferable to

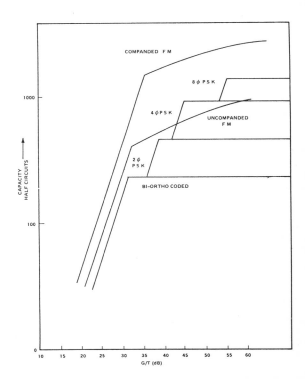

Fig. 1. Effect of Various FM/FDMA and PCM/PSK/TDMA Techniques on Satellite Transponder Capacity

FM/FDMA for the bandwidth limited class of satellites and that companded FM/FDMA gives greater capacity than the other alternatives considered. Upon experimental verification of these conclusions, an attractive alternative for certain applications of preassigned and demand-assigned circuits may be available.

Digital Modulation Techniques

Though frequency modulation is used almost exclusively in current satellite systems, digital modulation will be more widely used because of its increased bandwidth efficiency, increased capacity, reliability and flexible traffic-handling ability. It will permit large and small users equal access to the satellite system without sacrifice to either (8).

Digital modulation techniques for radio relay systems have been discussed in detail by Reinhart (9). Whelan (10) has presented a comparison of analog-FM and digital-PSK transmissions from spacecraft. Digital communication techniques for satellite communications have been the subject of a 1969 Intelsat/IEE conference (11) and a second conference held in Paris in 1972 (32). The third such international conference will be held in Kyoto, Japan in 1975.

The characteristics of digital communications that are advantageous in satellite communications are (12):

1. The capability of achieving any prescribed amount of error control in single, as well as cascaded, channels

2. Efficient tradeoff between bandwidth and signal-to-noise-ratio permitting processing gains higher than those obtainable with analog systems

3. Flexibility in message handling such as multiplexing signals with different characteristics, adding and dropping channels, routing, switching, storing and regenerating signals

4. Signal processing capability to give spectrum conservation through source encoding, redundancy reduction and data compression.

In addition to these general advantages, there are specific advantages that apply to improved utilization of the available RF spectrum and to advanced forms of satellite networking.

Bandwidth-efficient modulation techniques trade power for bandwidth; i.e., a given channel carrying a certain information rate uses more RF power but less RF bandwidth when a bandwidth-efficient modulation scheme is employed. Bandwidth efficiency is readily obtained in digital modulation systems by increasing the number of levels in a multi-level modulation scheme. Digital modulation schemes such as PSK also have the advantage of relatively low susceptibility to interference (as compared to analog FM) (13, 14).

Since consideration of satellite power is usually the first step in the selection of suitable modulation methods, it is necessary to adopt modulation techniques that involve signal-to-noise ratio improvement in the demodulation process (15). In general, the most effective digital methods in this

regard are 2- and 4-ary PSK techniques (9, 10, 16). While the increased resistance to interference provided by 2-ary PSK is attractive, the possibility of doubling the channel capacity in approximately the same bandwidth by using 4-ary PSK is well worth the cost (17). Figure 2 shows a comparison of digital systems based on average power. It is conceivable that future systems with both amplitude and phase modulation may be employed to provide even greater spectrum conservation (14, 18, 19).

Several areas in which digital techniques applicable to satellite communications are developing are the following:

1. MAT-1

Intelsat has developed a traffic handling system known as MAT-1, an experimental PCM/PSK/TDMA system, to satisfy the requirements of medium-density traffic networks whose links may consist of from 12 to several hundred channels (1). This 50-Mbit/s system has a capacity of over 700 voice channels that are distributed in multi-destinational time slots within repositionable bursts. As of mid-1974, several Intelsat members have issued procurement specifications for trial operation of TDMA systems.

2. SPADE

Terminals to operate in the SPADE (2, 20, 21) network with an Atlantic INTELSAT IV satellite have been installed and are in operation. SPADE is a demand assignment PCM/FDMA system which

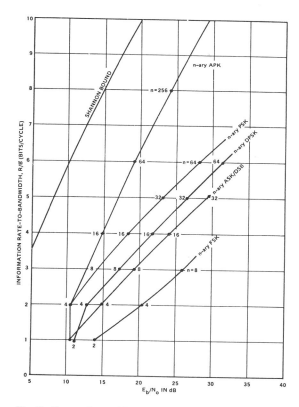

Fig. 2. Comparison of Digital Modulation Systems Based on Average Power

utilizes single-channel-per-carrier access to the satellite. In the operational configuration, the digital transmitted bit rate will be 64 kbit/s and the modulation used for each carrier will be four-phase CPSK.

Since SPADE is a demand assignment system it affords more efficient utilization of the satellite than is available using the present Intelsat FDM/FM operational system. The significant advantage of a demand-assignment multiple-access system is that it allows the sharing of satellite circuits by a large number of terrestrial users who are in common view of a satellite. The circuits are assigned on demand from a temporary connection between any two earth stations within the region covered by the satellite, and when no longer required, are returned to the demand-assignment pool.

Plans include the provision of wideband data-transmission (22–25) capability over an existing SPADE channel unit for such typical data rates as 48 kb/s and 56 kb/s. The wideband data stream can consist of wideband data received directly from the terrestrial link, multiplexed narrow-band data, or it can be formed by multiplexing telegraphy channels (25). Higher data-rate transmissions (1.554 Mb/s) are being considered in future applications.

3. DIGITAL TELEVISION

There has been interest in the digital transmission of television signals from satellites for program distribution and broadcast purposes because it offers long-haul transmission with significant reduction in the effects of distortion usually encountered in long-haul terrestrial radio-relay channels. Another feature of digital transmission which makes it attractive for the future is the ability to process the video signal to use the minimum required power and bandwidth needed to transmit the signal with a given fidelity. With the availability of high-speed digital circuitry it is now practical to implement bandwidth-reduction techniques such as frequency-interleaved sampling and frame replenishment (26–28). COMSAT has developed a digital television transmission system based on frequency-interleaved sampling that reduces the bandwidth occupancy by half for any particular digital modulation (28–31).

The British Broadcasting Corporation Research Department (14) has examined the possible advantages of using digital modulation for direct television broadcasting from space. It has concluded that a practical digital-modulation system, operating within the bandwidth that probably will be allocated to direct television broadcasting from satellites, is unlikely to permit a significantly lower transmitter power than that required for an FM system. However, digital systems may have an advantage of about 2 to 8 dB from the point of view of co-channel interference. A 20 PSK modulation is regarded as the best compromise between performance and complexity. The receiver adapter for a digital system is estimated to cost $52 more than the adapter for an FM system. BBC concludes that direct-broadcast systems are not likely to be adopted soon.

References

1. W.G. Schmidt, et al., "MAT-1: Intelsat's Experimental 700 Channel TDMA/DA System," *Intelsat/IEE Int. Conf. on Digital Satellite Communications,* London, 1969, Conf. Pub. #59, pp. 428–440.
2. A.M. Werth, "SPADE: A PCM FDMA Demand Assignment System for Satellite Communications," Ibid. London, November 1969, pp. 51–68.
3. B.G. Evans, and R. Walters, "An Economic Satellite Communications System for Small Nations," *IEEE Int. Conference on Communication Systems,* Montreal, June 1971, pp. 19–26.
4. D. Chakraborty, "Intelsat IV Satellite System (voice) Channel Capacity versus Earth-Station Performance," *IEEE Trans. Comm. Tech.* vol. COM-19, no. 3, pp. 355–362, June 1971.
5. D.R.C. Snowden, "A Small Station Satellite System Using Delta Modulation," *Australian Telecommunications Research ATR*, vol. 4, no. 1, pp. 3–9, May 1970.
6. D.I. Dalgleish, and A.G. Reed, "Some Comparisons of the Traffic Carrying Capacity of Satellites Using Digital Techniques with the Capacity of Satellites Using Frequency Modulation," *Intelsat/IEE Int. Conf. on Digital Satellite Comm.*, London, November 1969, pp. 226–240.
7. L.B. Dunn, "Telephony to Remote Communities in Canada via Satellite," *IEEE Int. Conference on Communications*, Montreal, 1971, pp. 11-20.25.
8. P.L. Bargellini, and W.G. Schmidt, "Digital Communications via Satellites: Advantages and Trends," *IEEE Electronics and Aerospace Systems Convention (EASCON) Record*, Washington, D.C., October 1970, pp. 203–208.
9. E.E. Reinhart, "Radio Relay System Performance in an Interference Environment," The Rand Corporation, Santa Monica, California, Memorandum RM-5786-NASA, October 1968.
10. J.W. Whelan, "Analogous-FM vs. Digital-PSK Transmission," *IEEE Trans. Comm. Tech.,* vol. COM-14, no. 3, pp. 275–282, June 1966.
11. *Intelsat/IEE International Conference on Digital Satellite Communications, Proceedings,* London, November 1969, p. 558.
12. J.V. Charyk, "Potentials of Digital Satellite Communication," Ibid.
13. "Effect of Modulation Techniques on the Re-use Density of the Geostationary Spectrum" in "Radio Spectrum Utilization in Space," A Report of the Joint Technical Advisory Council of the IEEE and EIA, The Institute of Electrical and Electronics Engineers, Inc., New York, September 1970, pp.53–75.
14. A. Brown, "Satellite Broadcasting: Possible Advantages of Using Digital Modulation for Television," BBC Research Report 1971/25, Research Department, Engineering Division, The British Broadcasting Corporation, Tadworth, Surrey (U.K.), July 1971.
15. S.G. Lutz, "Future Satellite Relayed Digital Multiple Access Systems," *Proc. Intelsat/IEE International Conference on Digital Satellite Communications,* London, November 1969, pp. 518–531.
16. H. Haberle, "Theoretical Comparison of Modulation Systems in Satellite Communications," *Nachrichtentechnische Zeitschrift-Communication Journal,* no. 3, p. 110, 1967.

17. L.C. Tillotson, "A Model of a Domestic Satellite Communications System," *Bell System Technical Journal*, pp. 2111–2137, December 1968.

18. W.L. Pritchard, "Communications Satellite Technology, Present and Future," Symposium on Long Term Prospects for Satellite Communications, Istituto Internazionale Delle Comunicazioni, Genoa, Italy, June 1971.

19. C.R. Cahn, "Combined Digital Phase and Amplitude Modulation Communication Systems," *IRE Trans. on Communication Systems,* vol. CS-8, pp. 150–155, 1960.

20. A.M. Werth, "SPADE: A PCM FDMA Demand Assignment System for Satellite Communications," *IEEE International Conference on Communications*, Paper no. 70-CP-423-COM, San Francisco, California, June 1970.

21. J.G. Puente, and A.M. Werth, "Demand Assignment Service for the Intelsat Global Network," *IEEE Spectrum*, pp. 59–69, January 1971.

22. B.I. Edelson, and A.M. Werth, "SPADE System Progress and Applications," *Conference on Ground Stations for Satellite Communication,* Tel Aviv, Israel, May 1971.

23. E.R. Cacciamani, "A Channel Unit for Digital Communications in the SPADE System," *IEEE International Conference on Communications Proc.*, Montreal, June 1971, pp. 42–15 to 42–19.

24. E.R. Cacciamani, "The SPADE Concept Applied to a Network of Large and Small Earth Stations," *AIAA 3rd Communications Satellite Systems Conference,* Paper No. 70–420, Los Angeles, April 1970.

25. E.R. Cacciamani, "Satellite Data Transmission Using SPADE," *EUROCON 71*, Lausanne, Switzerland, October 1971.

26. W.K. Pratt, "A Bibliography on Television Bandwidth Reduction Studies," *IEEE Trans. on Information Theory*, vol. IT-13, no. 1, pp. 114–115, January 1967.

27. A.J. Seyler, "The Coding of Visual Signals to Reduce Channel-Capacity Requirements," *IEE Monograph 535E*, London, July 1962, pp. 676–684.

28. A.K. Bhushan, "Efficient Transmission and Coding of Color Pictures," M.S. thesis, Department of Electrical Engineering, Massachusetts Institute of Technology, Cambridge, Mass., June 1967.

29. L.S. Golding, "Digital Television Transmission Systems for Satellite Communication Links," *1968 International Broadcasting Convention IEE Conf.* Publication no. 46, London, 1968.

30. L.S. Golding, "Frequency Interleaved Sampling of the NTSC Color Television Signal," *XVII Congresso Scientifico Internazionale per L'Elettronica*, Rome, March, 1971.

31. L.S. Golding, and R.K. Garlow, "Frequency Interleaved Sampling of a Color Television Signal," *IEEE Trans. on Communication Technology,* vol. COM-19, no. 6, pp. 972–979, December 1971.

32. The Second International Conference on Digital Telecommunications via Satellite, Paris, November 1972; Editions Chiron, Paris.

Frequencies Above 10 GHz

NELSON McAVOY
Goddard Space Flight Center
Greenbelt, Maryland

GENERAL DISCUSSION

The demand for satellite communications services is continually increasing and the capacity available within the 500-MHz frequency bands presently used at 4 and 6 GHz will not be sufficient to meet future needs. Therefore, additional spectrum space will be required with future systems and this can be provided by using frequencies above 10 GHz, where larger bandwidths are available.

The design of both satellites and earth stations in the higher frequency bands will not be significantly changed from those at lower frequencies and no major difficulties should arise due to the great amount of technology available concerning millimeter-wave circuitry. The main constraint on system design will be propagation factors above 10 GHz. To date estimates of propagation characteristics over satellite links have been based on limited data and it is this lack of knowledge which is holding back expansion into the frequency bands above 10 GHz.

The principal propagation limitation above 10 GHz occurs when precipitation intercepts the earth-space propagation path and causes attenuation and depolarization of the transmitted signal. It is necessary to find out the percentage of time that a given value of attenuation or depolarization is exceeded at the desired earth-station location. Little statistical data of this type are presently available, and thus, specification of path statistics is extremely difficult. The data which do exist allow order-of-magnitude estimates only to be made.

Clear-sky absorption due to gases in the atmosphere will be encountered in the higher frequency bands, but adequate data exist for predicting this (1, 2), and even though it varies with humidity and

elevation angle, an excess attenuation of only a few dB can be expected. Fortunately, some of the effects experienced in lower-frequency satellite communications such as ducting, ionospheric scintillation, and to a lesser extent, tropospheric scintillation, decrease with increasing frequency and are not troublesome to the proposed systems. Although time and frequency dispersions will limit the information bandwidth of the transmissions this has been shown not to be serious (3). Precipitation attenuation and the method of polarization rotation will have an impact on link power budgets. The method of polarization rotation will limit the achievable isolation between cross-polarized transmissions on the same frequencies.

PRECIPITATION ATTENUATION

With a knowledge of the shape of raindrops and of the distribution of drop sizes, calculations can be performed for attenuation. Based on spherical drops and the Laws and Parsons distribution, calculations have been performed by Medhurst (4), Oguchi (5) and Ryde and Ryde (6) giving attenuation for various rain rates at different frequencies. Such theoretical data can be used only for scaling measured data and not directly without making assumptions concerning the spatial distribution of the rainfall.

For efficient communications system design, the percentage of time during which a given value of precipitation is exceeded over the propagation path of interest must be known. The most reliable method for obtaining such a statistic is to measure the attenuation over a long period of time on a transmission path similar to the planned path. Unfortunately little data of this type exist and one usually has to resort to rainfall data in order to predict the attenuation. This process has severe limitations which are inherent in the methods used for collection of such data. In general instantaneous rain rate data are not available and only recently (7) have fast-acting gauges been developed. The other limitation lies in the spatial variation which depends greatly on the type of weather. Exact spatial and temporal characteristics of rainfall are still a matter of research. Hogg (8) has provided some of the answers in showing that the measurement of the intensity and spatial movement of rain cells is greatly affected by the sampling intervals of the rain gauges.

The experimental data obtained so far has mainly covered measurement of attenuation of the noise power from the sun (radiometer) or the sky temperature (9, 10, 11). This has been supplemented recently by actual measurement of attenuation through the atmosphere at 15.3 and 31.65 GHz for various locations in the U.S., using the ATS-5 satellite (12). Data from the referenced sources were taken by different methods using variations in period, frequency, location and elevation. Although the data can be transformed using empirical data between elevations, frequencies, and times, the process must remain suspect. However, Figure 3 shows an attempt to assign an order of magnitude to the problem for planning purposes at 12 GHz.

The distance traversed by an earth-space path in that part of the atmosphere in which rain can occur is a function of elevation angle. At very low angles, the path may encounter several convective cells, or it may traverse many miles of non-convective rain. At high angles, it is unlikely that the path will encounter more than one convective cell, and the distance through non-convective rain is less. However, it is not necessarily true that exposure continues to diminish with increasing elevation angle, because the worst exposure of all might be looking straight up into a convective cell.

It is generally assumed that a margin of 10 dB for rain attenuation is technically and economically feasible. Figure 3 indicates that a reliable service (outage less than 0.01 per cent) is feasible in all locations except the eastern seaboard of the U.S., where other techniques may be required (see space diversity).

The foregoing shows that the prevailing climate, specifically the rain-rate distribution, has a significant influence on the attenuation statistics along a particular transmission path. There are two basically different types of precipitation that must be considered: widespread precipitation at moderate rates which gives the low values of attenuation, and high-intensity, localized precipitation at highly variable rain rates which gives the high values of attenuation.

An indication of the widespread type of precipitation is the average annual rainfall rates which are available from meteorological sources and which correlate well with lower attenuation statistics. However, meteorologists are still using slow-acting rain gauges which are good enough for measuring widespread moderate rainfall but which do not measure heavy rainfall with sufficient accuracy. It

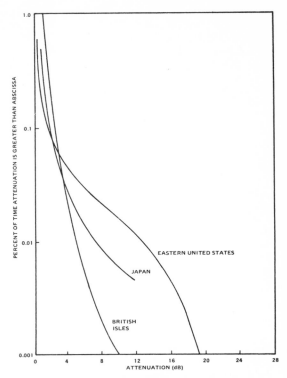

Fig. 3. World Attenuation Statistics at 12 GHz for Earth-Space Paths

is possible to see some correlation with charts of annual thunderstorm days, these being associated with the heavy convective rain, and the instances of high attenuation. To increase the accuracy of the estimate it is helpful to have rain-rate statistics for the site in question taken from a fast-acting rain gauge. These statistics show that the locations with high attenuation have correspondingly high rain rates.

Ippolito (13) found from measurements with ATS-5 that there is very poor correlation between estimated attenuation from point rate data and measured path attenuation. Improved correlation was obtained by using the average rainfall rate measured by rain gauges spaced evenly on the ground along the direction of the path, and an additional improvement was obtained by height-averaging the rainfall rate. However, measurement of rain attenuation by a sun tracker has resulted in good agreement with theory using the Laws and Parsons raindrop distribution.

SPACE DIVERSITY

One method of effectively reducing the large margins required for extreme rainfall rates occurring small percentages of the time is to provide space diversity on satellite paths or parallel-path diversity on terrestrial paths. For space diversity, earth stations separated at some distance, and for terrestrial links, parallel transmit-receive paths spaced at a distance are much less likely to have their paths intercepted by intense rain cells simultaneously. For space paths, space diversity measurements have been performed by Bell Labs on their sun tracker and radiometer experiments (14).

Space diversity can provide significant improvement for short periods of time with high rain rates (caused by local rain cells covering a small area) as can be seen from Figure 4. This data was obtained at 16 GHz but has been scaled to 12 GHz. If two earth stations are separated by ten kilometers (6 miles) the figure indicates that a rain margin of 2 dB at each station would be adequate to assure less than one hour per year of outage.

Rain attenuation increases with frequency. Theoretical analysis indicates that the attenuation values of Figure 3 would be increased by factors of about 2.5 and 4.5 at frequencies of 20 and 30 GHz. Therefore, at these higher frequencies, it appears that diversity would be necessary in most locations.

For large percentages of the time rain attenuation is due to low rainfall rates and clouds which cover a large area, and this could limit diversity usage as shown by Evans (15) for the New Jersey area. The latter analyzes the results of the BTL sun tracker and radiometer measurements. More information regarding the advantages of space diversity is being collected in the U.S. as part of the ATS program. Information on the European sector will become available from the SIRIO satellite experiment (16).

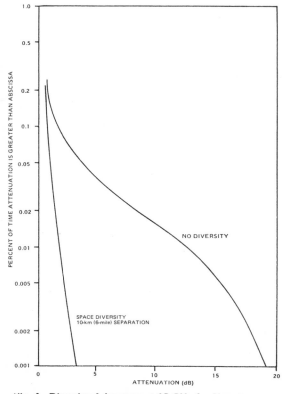

Fig. 4. Diversity Advantage at 12 GHz for New Jersey

POLARIZATION ROTATION DUE TO RAIN

The capacity of a communications system using a fixed frequency spectrum may be doubled by frequency reuse that employs the method of radiating on orthogonal polarizations. Whether this potential increase in capacity can be realized depends on the amount of cross-polarization discrimination that can be achieved for the co-channel operation. Cross-polarization discrimination is a function of the quality of the antenna systems and of the effect of the propagation medium on the polarization of the transmitted wave. This section is concerned with the polarization rotation effects of propagation through the atmosphere.

For a satellite system operating at centimeter and millimeter wavelengths, the primary polarization rotation is caused by clouds and rain intercepting the propagation path. A polarized wave incident on a medium composed of absorbing or scattering particles suffers both attenuation and polarization rotation while passing through the medium. If the particles are all spherical, and if single scattering is assumed, the wave which emerges from the medium is attenuated, but its polarization is not altered. On the other hand, if non-spherical particles and multiple scattering are assumed, then the wave is attenuated and its polarization changed. In most theoretical treatments of attenuation and scattering by raindrops, multiple scattering is not included. This omission may cause the validity of the results to be somewhat suspect, since for nearly all communications cases multiple scattering is important; however, it is difficult to account for in theory.

Cross-polarization is caused by anisotropy in the transmission medium. In the case of non-spherical raindrops one component derives from the differential attenuation along minor and major axes and the other from scattering (or the induction of currents and their reradiation by the drops). The effects of single non-spherical raindrops have been computed by Oguchi (6). However, information on the raindrop distribution is refined to apply Oguchi's results. Oguchi assumes ray-path alignment parallel to the surface of the earth. Information for terrestrial paths is scarce. For the nearly vertical earth-space paths such information is almost non-existent. Thomas (17) and Saunders (18) have presented data for the latter case indicating average-drop tilt angle (canting angle) of 15°. The cross-polarization discrimination versus rain rate as calculated by Saunders for a terrestrial link is given in Figure 5. Drops with positive and negative tilts cancel out and only a small proportion of the raindrops contribute to the cross polarization. A completely random orientation would give zero cross polarization. Results from slant-path measurements made from radar backscatter by Hodson and Peter (19) show an average peak in distribution around 10 to 15°, and thus there seems to be justification for assuming a raindrop orientation tendency. Furthermore, extrapolation to the space path from terrestrial data would give misleading results since the two paths have different weather.

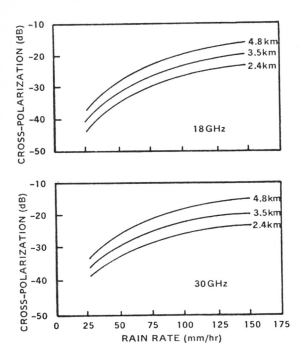

Fig. 5. Cross-Polarization Versus Rain Rate at 18 and 30 GHz [Path Length a Parameter (Saunders)]

The slant path, for instance, passes through the radar melting band known for its depolarization effects and transgresses regions of drop breakup also known to give high cross-polar values.

In comparing linear and circular polarizations, the worst case in linear polarization occurs for a raindrop oriented at 45° to the linear vector, and the same is true for circular polarization. The worst case for linear, then, is as good as any case for circular, and therefore linear should always be preferable.

It is possible to use Oguchi's results to obtain conservative estimates of cross polarization obtainable on the space path within the limitations of this technique. Cross-polarization discrimination at various frequencies is compared with total path loss for the maximum rain rate applicable under Oguchi's theory. For a 10-dB margin of attenuation, this indicates a discrimination of better than 24 dB at 12 GHz.

COMMANDABLE HIGH-POWER SPOT BEAMS

An alternative to space diversity as a method of overcoming the attenuation caused by precipitation is to increase by command the power of one or more of the spot beams on a satellite downlink. This technique is based on the unlikelihood of a number of beams being simultaneously subjected to precipitation attenuation.

Consider a satellite with 10 spot beams that is designed so that any two beams could be increased in power by 6 dB at the same time. Such a satel-

lite could counteract 6 dB precipitation attenuation occurring in the beams directed at two different geographical areas. The satellite would require only a 2 dB increase in prime power. This technique has been discussed by Kiesling and Meyerhoff (20). The correlation of precipitation attenuation over large geographical areas may be investigated by NASA as part of the feasibility assessment for this approach.

OPTICAL COMMUNICATIONS

The NASA program in optical communications is devoted to the use of CO_2 laser heterodyne systems for space-to-space links. These links include the data-relay link between a low-altitude earth-sensing spacecraft and a synchronous data-relay spacecraft, as well as the link between two synchronous data-relay satellites. The ten-year technology program (21–26) will culminate in late 1975 with the completion of the development of a 10.6-micrometer laser transceiver capable of transmitting or receiving data at rates in excess of 300 Mbit/s. The receiver accommodates doppler shifts of ±700 MHz, which meets the requirements for the low-altitude to synchronous-satellite link. The transceiver uses a circular (19-cm diameter) optical antenna having 93 dB of gain. The transceiver engineering model is 63.5 cm × 26.6 cm × 25.4 cm (25 in × 10.5 in × 10 in), weighs less than 45.4 kg (100 lbs), and requires less than 150 W of prime power.

The present engineering model uses a coupling-modulated laser (27–29) and a modulation format based on the Miller code (30). Miller coders and decoders have been built and tested at the 100, 200, and 300 Mb/s data rates and all perform within about 2 dB of their theoretical values. The laser modulation is analogous to the double-side-band, suppressed-carrier modulation of the RF spectrum and requires, therefore, carrier reconstruction in the receiver. This is accomplished by a squaring loop in the receiver's doppler tracking system (31). The performance of the squaring loop is also within about 2 dB of theoretical. The receiver front-end uses a waveguide laser local oscillator (32, 33) to establish the first intermediate frequency and a mercury-cadmium-telluride mixer (34, 35) which is cooled to 120 K by means of a radiation cooler (36). The front-end sensitivity is 10^{-19} W/Hz (the theoretical quantum limit is 1.88×10^{-20} W/Hz).

Continuing research work is devoted to the investigation of m-ary modulation formats (band-width conserving) which will permit the system to operate at data rates in excess of 1 Gb/s (37, 38).

While the work is directed towards NASA applications, it also has commercial communications satellite applications. The lightweight, low-power laser systems can add great flexibility to domestic and international communications satellite networks. Satellite-to-satellite links can be used to permit two or more satellites to operate as a single system for such applications as school video distribution and air traffic control. By eliminating the need to transfer data between satellites through a ground station, the laser systems conserve satellite-to-ground frequencies—a particular advantage in commercial applications (39). Other applications of satellite-to-satellite links include improvement of fringe area communications (such as Alaska and New England in present systems) and marine communications. These advantages were realized early in the NASA program (22–24) and, in spite of some of the setbacks in the NASA communications program, have remained a chief goal of that program following closely behind the support of the earth-sensing missions.

References

1. J.H. Van Vleck, "Absorption of Microwaves by Oxygen," *Phys. Rev.,* pp. 413–424, 1947.
2. J.H. Van Vleck, "Absorption of Microwaves by Uncondensed Water Vapour," *Phys. Rev.,* pp. 425–433, 1947.
3. R.K. Crane, "Coherent Pulse Transmission through Rain," *IEEE Trans. on Ant. and Prop.,* vol. AP-15, no. 2, pp. 252–256, March 1967.
4. R.G. Medhurst, "Rainfall Attenuation of Centimeter Waves: Comparison of Theory and Measurement," *IEEE Trans. on Ant. and Prop.,* vol. AP-13, pp. 550–564, July 1965.
5. T. Oguchi, "Attenuation of Electromagnetic Waves due to Rain with Distorted Raindrops, (Part II)," *Journal of the Radio Research Lab.,* vol. 11, no. 53, January 1964.
6. J.W. Ryde, and D. Ryde, "Attenuation of Centimetre and Millimetre Waves by Rain, Hail, Fogs and Clouds," Report No. 8670, G.E.C., May 1945.
7. J.R. Norbury, and W.J. White, "A Rapid-Response Rain Gauge," *J. Phys. E.,* vol. 48, no. 8, pp. 601–602, August 1971.
8. D.C. Hogg, "Path Diversity in Propagation of Millimetre Waves through Rain," *IEEE Trans. on Ant. and Prop.,* vol. AP-15, pp. 410–415, May 1967.
9. G.E. Mueller, "Propagation of 6-Millimetre Waves," *Proc. IRE,* no. 34, pp. 181–183, 1946.
10. Japanese Contribution "Method for Predicting the Long Term Statistics of Attenuation due to Precipitation of Radio Waves in the 10 GHz Band at Higher Angles of Elevation in Japan," CCIR Study Group 5, I.W.P., Nice, December 1970.
11. U.K. Contribution "Summary of U.K. Measurements up to 31 October 1970 on Slant Path Attenuation and Atmospheric Noise," CCIR Study Group 5, I.W.P., Nice, December 1970.
12. "Millimetre-Wave Propagation Experiments Utilizing the ATS-5 Satellite," NASA-X-751-70-428, November 1970.
13. L.J. Ippolito, "Effects of Precipitation on 15.3 and 31.65 GHz Earth-Space Transmissions with the ATS-V Satellite," *Proc. IEEE,* pp. 189–205, February 1971.
14. R.W. Wilson, "Attenuation on an Earth-Space Path Measured in the Wavelength Range of 8 to 14 Micrometres," *Science,* 168, pp. 1456–57, June 19, 1970. "A Three-Radiometer Path Diversity Experiment," *Bell System Tech. Jour.,* pp. 1239–42, July 1970.
15. H.W. Evans, "Attenuation on Earth-Space Paths at Frequencies up to 30 GHz," *IEEE Int. Comm. Conf.,* Montreal, 1971, pp. 27.1–27.5.
16. F. Carana, G. Drufuca, and A. Paraboni, "The Italian Satellite SIRIO GHF Propagation and Com-

munication Experiment," *AIAA 3rd Comm. Satellite System Conf.,* Los Angeles, April 1971.
17. D.T. Thomas, "Cross-Polarization of Microwave Transmission due to Rain," presented at USNC/URSI Fall meeting, Columbus, Ohio, 1970.
18. M.J. Saunders, "Cross-Polarization at 18 and 30 GHz," *IEEE Trans. on Ant. and Prop.,* vol. AP-19, no 2, pp. 273–277, March 1971.
19. M. Hodson, and P. Peter, "Observations of the Ellipticity of Raindrops Using a Polarized Radar System," *World Conference on Radio Meteorology,* Boulder, Colorado, September 1964.
20. J.D. Kiesling, and H.J. Meyerhoff, "TV Satellite Distribution at Frequencies above 10 GHz-Communication Satellites for the 70's: Systems," *MIT Press,* vol. 26, p. 171, 1971.
21. N. McAvoy, "10.6 Micron Communication Systems," NASA TM X-524-65-461, Goddard Space Flight Center, November 1965.
22. N. McAvoy, H.L. Richard, J.H. McElroy, and W.E. Richards, "10.6-Micron Laser Communications System Experiment for ATS-F and ATS-G," NASA TM X-524-68-208, Goddard Space Flight Center, May 1968.
23. J.H. McElroy, N. McAvoy, H.L. Richard, W.E. Richards, and S.C. Flagiello, "An Advanced 10.6 Micron Laser Communication Experiment," NASA TM X-524-68-478, Goddard Space Flight Center, November 1968.
24. J.H. McElroy, "Carbon Dioxide Laser Systems for Space Communications," *Proc. IEEE 1970 International Conference on Communications,* pp. 22–27.
25. Laser Technology Branch, "NASA Laser Data Relay Link (LDRL) Experiment for the DOD/NASA Cooperative Space Laser Communication Test Flight," Goddard Space Flight Center, May 1974.
26. J.H. McElroy, "Carbon Dioxide Laser Space Data Relay Links," Paper MA2, Presented at 1974 Optical Society of America, Spring Meeting.
27. N. McAvoy, J. Osmundson, and G. Schiffner, "Broadband CO_2 Laser Coupling Modulation," *Applied Optics,* vol. 11, no. 2, p. 473, February 1972.
28. J.E. Kiefer, T.A. Nussmeier, and F.E. Goodwin, "Intracavity CdTe Modulators for CO_2 Lasers," *IEEE J. Quantum Electron.,* vol. QE-8, no. 2, pp. 173–179, February 1972.
29. D.R. Hall, C.J. Peruso, E.H. Johnson, G. Schiffner, J. McElroy and N. McAvoy, "Multichannel Television Coupling Modulation Experiments Using a CO_2 Laser," *Optical Engineering,* vol. 11, no. 3, pp. 77–82, May/June 1972.
30. M. Hecht, and A. Guida, "Delay Modulation," *Proc. IEEE,* vol. 57, no. 7, pp. 1314–1316, July 1969.
31. T. Flattau, and J. Mellars, "Wideband Infrared Receiver Backend," Final Report NASA Contract NAS5-23183, April 1974.
32. J.J. Degnan, and D.R. Hall, "Finite Aperture Waveguide Laser Resonators," *IEEE J. Quantum Electron.,* vol. QE-9, no. 9, pp. 901–910, September 1973.
33. R.L. Abrams, "Gigahertz Tunable Waveguide CO_2 Laser," *Appl. Phys. Ltrs.,* vol. 25, no. 5, pp. 304–306, 1 September 1974.
34. B.J. Peyton, A. DiNardo, R. Kanishak, and F.R. Arams, "High Sensitivity Receiver for Infrared Laser Communications," *IEEE J. Quantum Electron.,* vol. QE-8, no. 2, pp. 252–263, February 1972.
35. B.J. Peyton, "Wideband Infrared Heterodyne Receiver Front-End," Final Report NASA Contract NAS5-23119, August 1974.
36. F. Gabron, J.E. McCullough, R.L. Merriam, "Spaceborne Passive Radiator for Detector Cooling," *Proc.*

20th National Infrared Information Symposium (IRIS), 1972.

37. P.F. Panter, *Modulation, Noise, and Spectral Analysis,* McGraw-Hill, 1965, ch. 20, 21.

38. Technical Staff, Bell Telephone Laboratories, *Transmission Systems for Communications:* BTL (Revised Fourth Edition, 1971), p. 629.

39. J.E. Burtt, C.R. Moe, R.V. Elms, L.A. DeLateur, W.E. Sedlacek, and G.G. Younger, "Technology Requirements for Communication Satellites in the 1980's," Final Report NASA Contract NAS2-7073, September 1973.

Techniques for Frequency Reuse

Extracted From
"A Review of Satellite
Systems Technology" (1972)
[IEEE 72CHO749-2 AES]

The capacity of bandwidth-limited communications satellites may be increased by utilizing frequencies more than once. There are two ways of avoiding interference in a channel that is used more than once: (1) by orthogonal polarization and (2) by multiple-exclusive spot beams.

ORTHOGONAL POLARIZATION

The use of opposite-hand circular or crossed-linear polarizations may be used to effect an increase in the bandwidth by a factor of two. Whether this potential increase in capacity can be realized depends on the amount of cross-polarization discrimination that can be achieved for the co-channel operation. Cross-polarization discrimination is a function of the quality of the antenna systems and of the effect of the propagation medium on the polarization of the transmitted signal.

Propagation Constraints

Estimates of cross-polarization discrimination for satellite links may be made in the 4–6 GHz bands. For higher frequency bands more measured data are needed. If 25 dB is an acceptable objective for the modulation system used, indications are that this can be met for a path fade of less than 10 dB. However, the only results available on path diversity techniques are for terrestrial links and these cannot be applied to the slant path with any degree of confidence. Some statistical data is required for various geographical locations before any systems planning along these lines can be accomplished.

Straight-up paths should give no polarization rotation, but low-elevation paths could give excessive rotation. Also, the ellipticity of the spacecraft antenna near its beam edges introduces extra cross-polarization terms.

Terminal Equipment

Consideration of the propagation medium indicates that linear orthogonality would give better performance than circular. This implies that the satellite stabilization needed for linear polarization would need to be precise to minutes of arc and the earth station would be required to track the polarization. Circular polarization, although inherently inferior, would not require such sophistication. The antennas themselves would determine the cross-polarization ratio.

Antennas

The polarization design problems are different for spacecraft and earth-station antennas, mainly due to the different coverages.

Spacecraft

The coverage of the spacecraft antenna varies from a small spot (1°) to full global coverage (17°), and in each case the polarization must be preserved over the field of view. Separating the transmit and receive antennas on the spacecraft eases the design problems considerably since the bandwidth is thereby reduced.

In general, horn-type radiators are used for global coverage and reflector-type antennas for spot beams. The requirements for both types of antenna are similar for dual-polarized operation since both feeds must have good cross-polar performance. The requirements are a small axial ratio (0.5 dB), a well defined phase center, maximum gain compatible with the beamwidth and, for circular polarization, identical orthogonal radiation patterns.

By far the most promising device appears to be the corrugated horn (1) whose construction ensures axial symmetry. This device is suitable for both linear and circular polarization, and isolation on the order of 40 dB is readily achievable. Multimode horns (2) could be an alternative, perhaps in the form of a main horn surrounded by a cluster of smaller horns, but again the critical factor is the multi-moding in the center main horn to produce low cross-polarization.

For the spot-beam antenna the polarization properties of the system depend on the reflector curvature (its f/D ratio) as well as the feedhorn characteristics. To realize good polarization properties from the reflector system, the feed must generate currents on the reflector which radiate fields into the aperture plane of the reflector that are spatially parallel. In a dual-polarized system the feed must be capable of achieving these conditions for both polarizations. For circular polarization the additional requirement of axial amplitude symmetry is imposed. The proportion of cross-polarization produced by the reflector depends on the f/D ratio. In general the larger the f/D ratio the more nearly parallel are the currents induced on the reflector. However, reduction of cross-polarization by manipulating the f/D ratio necessitates small-angle illumination and the associated problems of increased noise and reduced efficiency.

The introduction of a second reflector as in a cassegrain complicates the situation still further (3). However, in a cassegrain with a dual-mode feed and a high subdish magnification, the f/D ratio does not affect the cross-polarization discrimination, and this can be made large.

The constraints imposed on the feed system by the requirements of zero or negligible cross-polarization in the aperture plane indicate that, for feeds with axial symmetry, the primary feed pattern must be axially symmetric (i.e., its E and H plane patterns identical). This infers that the polarization characteristics of the feedhorn are similar to those required for the global-coverage horns. Thus hybrid-mode horns such as the corrugated and dual-mode (TE_{11} and TM_{11}) horns should provide good characteristics and 40 to 50 dB isolation should be achievable with axial ratios less than 0.5 dB.

For a linearly polarized system, the feed design problem can be alleviated by designing the reflector as a grid of parallel wires or conducting strips or by using a spatial polarization filter (4). In effect, this approach forces the aperture fields to be parallel independently of the feed characteristic. Using such a design it is possible to achieve isolation of 50 dB but this is not readily applicable to the circular case.

Earth Station

Since the earth-station beam is narrow and looks only at a single satellite, the polarization characteristics are of primary concern only along the beam axis and may be slightly affected by tracking and pointing errors. The reduction in antenna beamwidth considerably relaxes the design requirements. The off-axis cross-polarization sidelobes must, however, be low to avoid carrier degradation by pick-up from other systems.

The elements which influence the polarization properties of the station are the feedhorn and its compatibility with the reflector system, and the polarization diplexing arrangement.

The feedhorn design for both linear and circular polarizations is well developed and follows the basic lines already discussed for the satellite spot-beam antenna. For linear polarization, however, rotation control must be provided to allow for alignment with the satellite polarization; this is very critical (a few minutes of arc) for good cross-polarization discrimination. This control may be accomplished by mechanically rotating all or part of the feed system. An added complication may be that, because of the anisotropic nature of the propagation medium, independent rotation for the transmit and receive functions must be provided, and this would require a dual system. In the case of a single dish with dual feed, in order that the apparent feed center for both feeds be at the same point, it is necessary for at least one of the feeds to be in the form of an array. The alternative is to have two separate dishes with independent feeds.

The ease with which tracking may be incorporated must be considered in the choice of feed construction. For the case of single-horn feeds, tracking may adequately be performed by extracting higher-order modes which are induced in the horn when the satellite is offset from the horn axis. The waveguide arrangement required for this purpose introduces another source of antenna cross-polarization, but by careful design this element may be minimized.

The main problem as far as the earth-station equipment is concerned is that of separating two orthogonally polarized signals received from an antenna feed and combining two high-power polarized signals into an orthogonal configuration for transmission to the feed. This may be achieved by various systems involving polarization couplers, circulators, polarizers and either single- or dual-mode feeds. All of these will require the development of a polarization coupler with an isolation better than that produced by the dish (i.e., 40-50 dB) over both the transmit and receive bandwidths. As stated earlier, on-axis cross-polarization response is the main concern in auto-tracked ground-station antennas, and this is determined primarily by the isolation inherent in the diplexing equipment over the frequency range of interest. In the case of dual-feed systems, especially in cassegrain configurations, it is determined by any coupling between the feedhorns. The latter problems would be considerably simplified if separate dishes were used, but this may not be economically feasible.

For the earth-station antenna, precipitation on the device itself can cause severe degradation of its polarization performance (5). This would indicate some form of protection for the antenna in the form of a radome. The most severe depolarization occurs with water on the feed and subreflector and relatively little on the main dish itself. This effect, coupled with the attenuation effects of large radomes, suggests protection for the feed and subreflector only. With a small radome, water film on the radome would still be troublesome and it may be necessary to use some form of forced-air protection for these critical areas.

The choice between linear and circular polarization is still unresolved. From the propagation standpoint, linear is the best choice, but in practice has certain drawbacks. Figures 6 and 7 show the one-way polarization isolation for both circular and linear polarization as a function of the terminal antenna performance only. For well designed antennas a circular axial ratio of 0.5 dB should be obtainable at both ends. This corresponds to a cross-polar discrimination of 25 dB. The linear polarization case shown in Figure 7 shows link isolation against axial ratio of the spacecraft antenna (polarization shown in brackets). A well designed earth station should show a 40 dB discrimination and the axial ratio of the spacecraft antenna should be approximately 0.5 dB. From this figure it is seen that this corresponds to a link isolation of between 30 and 35 dB.

The foregoing would seem to indicate the superiority of linear polarization. However, Figure 8 shows how the isolation varies with misalignment of the antennas for linear polarization. Hence in order to realize the advantages of linear polarization extremely accurate satellite stabilization is required and possibly the complexity of polarization tracking. On the other hand, if orthogonal-

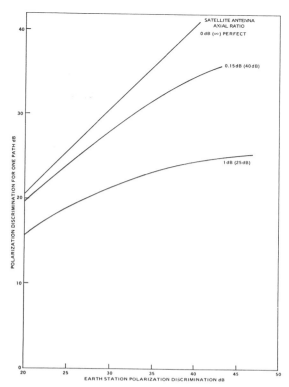

Fig. 7. Polarization Discrimination for Satellite-to-Earth Station Link Due to Antennas Alone (Linear Polarization)

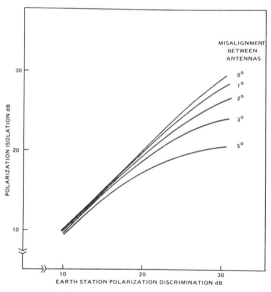

Fig. 8. Polarization Isolation for Satellite-to-Earth Station Link (Linear Polarization) Against Misalignment

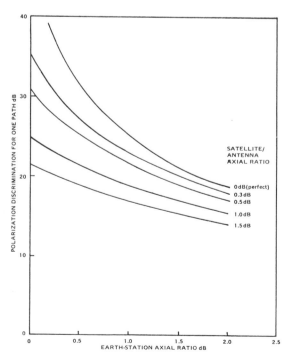

Fig. 6. Polarization Discrimination for Satellite-to-Earth Station Link Due to Antennas Alone (Circular Polarization)

polarization frequency reuse were considered at lower frequencies (say in the 4- and 6-GHz bands) the immunity of circular polarization to ionospheric rotations would be a major consideration.

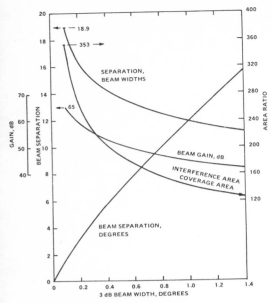

Fig. 9. Relations for Multi-Beam Satellites

MULTIPLE EARTHWARD "SPOT" BEAMS WITH FREQUENCY REUSE

It has long been recognized that a greater reuse of the frequencies would become possible whenever satellites could have narrow antenna beams, capable of covering small areas of earth (4,6,7).

This technique would permit closer spacing of co-frequency satellites and the use of multi-beam satellites. The relationships are shown in Figure 9.

References

1. P.J.B. Clarricoates and P.K. Saha, "Scalar Feeds for Earth Station Antennas," *IEE Con. Earth Station Technology,* Publication No. 72, 1975.
2. G. De Vito, "A Feed System for Satellite Communication Earth Stations," *Alta Frequenza,* vol. 39, no. 2, February 1970.
3. W.V.T. Rusch, "Phase Error and Associated Cross-Polarization Effects in Cassegrain-fed Microwave Antennas," Technical Report No. 32-610 T.P.L., Pasadena, California, May 1964, and *IEEE Trans. Ant. and Prop.,* May 1966.
4. R.L. Granger, "Polarization Diversity Study," Comsat Technical Memorandum CL-56-70, December 1970.
5. B.G. Evans, et al., "Investigation of the Effects of Precipitation on Parabolic Antennas Employing Linear Orthogonal Polarization at 11 GHz," *Electronics Letters,* vol. 7, no. 13, July 1971.
6. J.L. Hult, et al., "The Technology Potentials for Satellite Sapcing and Frequency Sharing," The Rand Corporation, Memorandum RM-5785-NASA, Santa Monica, California, October 1968.
7. M.C. Jeruchim, et al., "Orbit/Spectrum Utilization Study," G.E. Space Systems Org., Volume II-IV and Interim Report, Valley Forge Space Center, Philadelphia, Penna. 19101, 1969-1970.

Part 3

Frequency And Orbit Coordination And Utilization

Coordination of Geostationary Satellite Systems

DONALD JANSKY
Office of Telecommunications Policy
Washington, D.C., U.S.A.

INTRODUCTION

Increased use of the geostationary satellite orbit by satellites and associated earth stations may increase the probability of interference between satellite networks when common frequency bands are used. The number of parameters characterizing a system is so large that it is useful to devise a simple method to determine whether there is any risk of interference between two given satellite networks. The method described in the following is based on the concept that the noise temperature of the system receiving interference undergoes an apparent increase due to the effect of the interference. It can therefore be used irrespective of the modulation characteristics of the satellite networks concerned and the exact frequencies employed.

In this method, the apparent increase in the equivalent satellite-link noise temperature resulting from interference caused by a given system is calculated and this value is compared with a predetermined increase in the noise temperature.

CALCULATION OF THE INCREASE IN NOISE TEMPERATURE OF THE SATELLITE LINK RECEIVING INTERFERENCE

This analysis is performed as follows (1, 2):

Let A be a satellite link of network R associated with satellite S and A′ be a satellite link of network R′ associated with satellite S′. The symbols such as a, b, and c refer to satellite link A and symbols such as a′, b′ and c′, refer to satellite link A′.

The parameters are defined as follows (for satellite link A):

ΔT_s = increase in the receiver noise temperature of the satellite S caused by interference in the receiver of this satellite (K);

ΔT_e = increase in the receiver noise temperature of the earth station or caused by interference in the receiver of this station (K);

p_s = maximum power density per Hz delivered to the antenna of satellite S (averaged over the worst 4 kHz band for a carrier frequency below 15 GHz or over the worst 1 MHz band above 15 GHz) (W/Hz);

$g_3(\eta)$ = transmitting antenna gain of satellite S in the direction η (numerical power ratio);

η_A = direction, from satellite S, of the receiving earth station e_R of satellite link A;

$\eta_{e'}$ = direction, from satellite S, of the receiving earth station $e_{R'}$ of satellite link A′;
Note.—the product $p_s g_3 (\eta_{e'})$ is the maximum e.i.r.p. per Hz of satellite S in the direction of the receiving earth station e_R' of satellite link A′;

$\eta_{s'}$ = direction, from satellite S, of satellite S′;

p_e = maximum power density per Hz delivered to the antenna of the transmitting earth station e_T (averaged over the worst 4 kHz band for a carrier frequency below 15 GHz or over the worst 1 MHz band above 15 GHz) (W/Hz);

$g_2(\delta)$ = receiving antenna gain of satellite S in the direction δ (numerical power ratio);

δ_A = direction, from satellite S, of the transmitting earth station e_T of satellite link A;

$\Delta_{e'}$ = direction, from satellite S, of the transmitting earth station e'_T of satellite link A′;

$\delta_{s'}$ = direction, from satellite S, of satellite S′;

$g_1(\theta)$ = transmitting antenna gain of the transmitting earth station e_T in the direction of satellite S (numerical power ratio);

$g'_1(\theta)$ = transmitting antenna gain of the earth station e_T in the direction of satellite S′ (numerical power ratio);

$g_4(\theta)$ = receiving antenna gain of the earth station e_R in the direction of satellite S (numerical power ratio);

$g'_4(\theta)$ = receiving antenna gain of the earth station e_R in the direction of satellite S′ (numerical power ratio);

k = Boltzmann's constant (J/K);

ℓ_d = free-space transmission loss on the down-path (numerical power ratio)*;

ℓ_u = free-space transmission loss on the up-path (numerical power ratio)*;

γ = transmission gain of the satellite link evaluated from the output of the receiving antenna of the space station S to the output of the receiving antenna of the earth station e_R (numerical power ratio, usually less than 1), given by

$$\gamma = \frac{p_s g_3(\eta_A) g_4(\theta)\ell_u}{p_e g_1(\theta) g_2(\delta_A)\ell_d} \qquad (1)$$

θ = geocentric angular separation between two satellites (degrees).

The parameters ΔT_s and ΔT_e are given by the following equations:

$$\Delta T_s = \frac{p'_e g'_1(\theta) g_2(\delta_{e'})}{k\ell_u} \qquad (2)$$

$$\Delta T_e = \frac{p'_s g'_3(\eta_e) g_4(\theta)}{k\ell_d} \qquad (3)$$

In the foregoing equations, the gains $g'_1(\theta)$ and $g'_4(\theta)$ are those of the earth stations concerned. In the event that precise numerical data relating to earth station antennas are not available, the reference radiation pattern given in draft Recommendation 465 (Rev. 72) should be used.

The symbol ΔT is used to denote the increase in the equivalent noise temperature for the entire satellite link at the receiver input of the receiving earth station e_R due to interference from network R′.

---

*To simplify the calculation it was assumed that: basic transmission loss on the down-path is the same regardless of the satellite and earth station considered; basic transmission loss on the up-path is the same regardless of the earth station and satellite considered.

This increase is the result of interference entering at both the satellite and earth station receivers of link A. When satellites S and S′ are equipped with simple frequency-changing repeaters having the same translation frequency, the interference received by link A is caused on the up path and down path by the same link A′.

This can therefore be expressed as follows:

$$\Delta T = \gamma \Delta T_s + \Delta T_e \qquad (4)$$

Hence

$$\Delta T = \gamma \frac{p'_e g'_1(\theta) g_2(\delta_{e'})}{k\ell_u} + \frac{p'_s g'_3(\eta_e) g_4(\theta)}{k\ell_d} \qquad (5)$$

Equation (5) combines both the up-path and the down-path interference.

When the translation frequencies of the two satellites are not the same, different links in network R′ may interfere with link A at the satellite and earth station receivers; let these links be called A′ and $\bar{A}'$ respectively (the parameters such as a′, b′ and c′ relate to link $\bar{A}'$). Then:

$$\Delta T = \gamma \frac{p'_e g'_1(\theta) g_2(\delta_{e'})}{k\ell_u} + \frac{\bar{p}_s \bar{g}_3(\eta_e) g_4(\theta)}{k\ell_d} \qquad (6)$$

If there is a change of modulation in the wanted satellite then it may be necessary to treat up and down paths separately using equations (2) and (3).

In the same way, the increase $\Delta T'$ in the equivalent noise temperature for the entire satellite link at the receiver input of the receiving earth station e'_R under the effect of the interference caused by network R is given by the following equations:

$$\Delta T'_{s'} = \gamma \frac{p_e g_1(\theta) g'_2(\delta_e)}{k\ell_u} \qquad (7)$$

$$\Delta T'_{e'} = \frac{p_s g_3(\eta_{e'}) g'_4(\theta)}{k\ell_d} \qquad (8)$$

When both satellites share the same translation frequency, then

$$\Delta T' = \gamma' \frac{p_e g_1(\theta) g'_2(\delta_e)}{k\ell_u} + \frac{p_s g_3(\eta_{e'}) g'_4(\theta)}{k\ell_d} \qquad (9)$$

When the two satellites have different translation frequencies (calling two links of the R network A and $\bar{A}$ and denoting the corresponding parameters $\bar{a}$, $\bar{b}$, and $\bar{c}$):

$$\Delta T' = \gamma' \frac{p_e g_1(\theta) g'_2(\delta_e)}{k\ell_u} + \frac{\bar{p}_s \bar{g}_3(\eta_{e'}) g'_4(\theta)}{k\ell_d} \qquad (10)$$

For the two multiple-access satellites this calculation must be made for each of the satellite links established via one satellite in relation to all of the satellite links established via the other satellite.

COMPARISON BETWEEN CALCULATED AND PREDETERMINED PERCENTAGE INCREASE IN EQUIVALENT SATELLITE LINK NOISE TEMPERATURE

The calculated values of ΔT and $\Delta T'$ shall be compared with the corresponding predetermined values. These predetermined values are taken as 2% of the appropriate equivalent satellite link noise temperatures (see Radio Regulations, Appendix 29):

– if the calculated value of ΔT is less than or equal to the predetermined value, the interference level from satellite link A' to satellite link A is permissible irrespective of the modulation characteristics of the two satellite links of the exact frequencies used;

– if the calculated value of ΔT is greater than the predetermined value, a detailed calculation shall be carried out.

The comparison of $\Delta T'$ with the predetermined value shall be carried out in a similar manner.

As an example, it can be seen that in the case of a satellite link operating in accordance with current CCIR Recommendations using FM telephony and having a total noise in a telephone channel of 10,000 pWOp including 1000 pWOp interference noise from terrestrial radio-relay systems and 1000 pWOp interference noise from other satellite links, a 2% increase in equivalent noise temperature would correspond to 160 pWOp of interference noise.

DETERMINATION OF THE SATELLITE LINKS TO BE CONSIDERED IN CALCULATING THE INCREASE IN EQUIVALENT SATELLITE LINK NOISE TEMPERATURE FROM THE DATA FURNISHED FOR THE ADVANCE PUBLICATION OF A SATELLITE NETWORK

The greatest increase in equivalent satellite link noise temperature induced in any link of another satellite network, existing or planned, by interference produced by the proposed satellite network, must be determined.

The most unfavorably sited transmitting earth station of the interfering satellite network should be determined for each satellite receiving antenna of the network suffering interference. This is done by superimposing the "earth-to-space" service areas of the interfering network on the space-station receiving-antenna gain contours plotted on a map. The most unfavorably sited transmitting earth station is the one in the direction of which the satellite receiving-antenna gain of the network interfered with is the greatest.

The most unfavorably sited receiving earth station of the network suffering interference should be determined in an analogous manner for each "space-to-earth" service area of that network. The most unfavorably sited receiving earth station is the one in the direction of which the satellite transmitting-antenna gain of the interfering network is the greatest.

When the satellite of the network suffering interference is equipped with simple frequency-translating transponders, the determinations are made in pairs, one for the receiving antenna of a particular transponder and one for the space-to-earth service area associated with the transmitting antenna of that transponder.

The calculation described may be used to determine the greatest increase in equivalent noise temperature caused to any satellite link in a proposed satellite network by interference from any other satellite network.

References

1. Radio Regulations, International Telecommunication Union, Appendix 29.
2. CCIR Report 454.

Factors Affecting Orbit Utilization

DONALD JANSKY
Office of Telecommunications Policy
Washington, D.C., U.S.A.

INTRODUCTION

This section provides a summary of the principles which govern the minimum separation needed to avoid interference between satellites employing common frequencies. Efficient use of the geostationary satellite orbit and the frequency spectrum depend on good management principles as well as technical matters. Most important of all, perhaps, is the effective use of the procedures established by the World Administrative Radio Conference for Space Telecommunications, 1971, for the coordination of frequencies assigned to space and earth stations. These procedures are contained in Articles 7 and 9A of the Radio Regulations. The material in the sections that follow is largely taken from CCIR Report #453 (Rev. 74).

EARTH STATION ANTENNA RADIATION PATTERN

The radiation pattern of the earth-station antenna, especially the first 10° from the principal axis in the direction of the geostationary satellite orbit, is important in determining the interference between systems using geostationary satellites. A reduction in sidelobe levels would increase significantly the efficiency of utilization of the geostationary satellite orbit.

Antenna patterns considerably better than the reference radiation pattern shown in Figure 1 (taken from CCIR Recommendation 465), may be achieved by careful control of the sidelobe levels. High sidelobe levels are caused mainly by scattering from blockage in the aperture of the antenna. Certain antennas have no such blockage and may be designed to reduce sidelobe radiation.

Orbit utilization efficiency is enhanced when systems using earth stations with both high gain and a high figure-of-merit (G/T) are involved.

POLARIZATION DISCRIMINATION

The use of orthogonal linear or circular polarizations permits discrimination between two emissions in the same frequency band, from the same satellite or from closely adjacent satellites. This augments discrimination provided by the directional properties of satellite and earth station antennas. A detailed discussion of this topic is to be found under Techniques for Frequency Reuse.

Until more information is obtained on the polarization discrimination achievable in the main beam of a variety of types of satellite and earth station antennas, and until a better understanding is obtained of the extent of irremovable propagation degradations of wave polarization, some doubt remains as to whether it will be possible to achieve the 20 to 30 dB of discrimination that is required for in-beam frequency reuse.

If adjacent single-polarization satellites use orthogonal linear or circular polarizations, it may be possible to use the polarization discrimination in the sidelobes of the earth station antennas to reduce interference between the satellite networks, and to allow satellite spacing to be reduced. The sidelobe polarization discrimination obtainable in this way would be small, although even a few decibels of discrimination would permit a significant reduction in satellite spacing. However, it would not be possible to realize even this benefit systematically without adopting preferred polarization characteristics. This would involve a choice between linear and circular polarization and, where linear polarization is adopted, a choice of the preferred plane of polarization. There is not, at present, sufficient information to allow such choices to be made.

Polarization discrimination, whichever form is used, may permit the reuse of frequency within

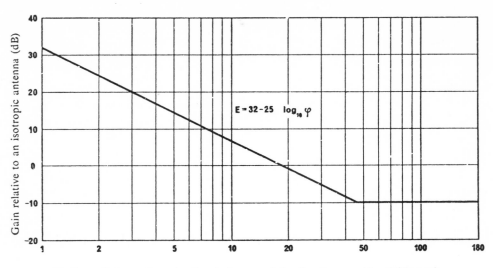

$$E = 32 - 25 \ \log_{10} \varphi$$

Angle, φ, between the axis of main beam and the direction considered (degrees)

Fig. 1. Provisional Reference Radiation Diagram

the same satellite beam. The following conclusions apply to possible increases in channel capacity, given the necessary wanted-to-unwanted carrier ratios for the overall link.

• For similar co-channel FDM/FM emissions, wanted-to-unwanted carrier ratios in the range 23–31 dB would be sufficient to limit mutual interference to 1000 pWp. Given that ratio, the total channel capacity for a bandwidth-limited system would be increased by 60% for no increase in the total downlink power, and by 100% given twice the total downlink power. There would be no increase in capacity if the satellite system were power-limited.

• For similar co-channel 4-phase PCM/PSK emissions, and for a bit error rate of 10^{-4}, a wanted-to-unwanted carrier ratio of about 20 dB and an increase in the total downlink power of 4 dB would permit the total channel capacity to be doubled. For 8-phase PCM/PSK emissions, the corresponding carrier ratio would be about 25 dB.

• By interleaving the carrier frequencies, the required wanted-to-unwanted carrier ratios for FDM/FM emissions would be reduced. No corresponding advantage is foreseen with digital transmission.

FREQUENCY REUSE

The use of the same carrier frequencies to serve different areas of earth by adjacent geostationary satellites can be greatly facilitated by the use of satellite antennas having an effective beamwidth much less than the angle which the earth subtends at the geostationary orbit. This means that the angular separation between the earthward beams is less than 17° and that, to obtain the necessary discrimination between the wanted signal and the unwanted signal, the mainlobe patterns of satellite antennas should conform to the coverage areas as closely as possible (by beam shaping in the plane normal to the direction of propagation). However, it will be necessary to define the concept of coverage area. In addition, beam shaping within the coverage area is desirable in order to maximize the satellite EIRP, particularly towards the earth stations in the coverage area. The discrimination offered by cross-polarization usually can be used to increase the effective sidelobe discrimination and thus can be employed to decrease the angular separation required between beams or to increase the possible beamwidths. Under certain circumstances, the same satellites may transmit separate information on the same frequencies two or more times over using antennas serving different parts of the world.

The radiation outside the coverage area of satellite antennas should be controlled. The utilization of techniques to reduce the first sidelobe level and to attenuate all sidelobes that illuminate the earth is to be encouraged. In this respect studies of satellite antenna types have shown lower sidelobe levels for those configurations which do not have blockage in the aperture. However, the design of complex systems without aperture blockage presents problems which are not yet fully understood. The radiation levels outside the desired coverage area also can be reduced by the use of large apertures with tapered illumination. However, the use of a large aperture raises spacecraft size and mass problems; furthermore, the use of tapered illumination shapes the main beam. This causes a rapid fall in gain at its edge, and so makes more stringent the requirements for satellite attitude and orbital control when area coverage is critical. (Beam shaping may reduce the required precision of attitude and orbital control in some cases; e.g., when the coverage area is unavoidably greater than the service area.)

To facilitate studies on the reuse of spectrum by narrow satellite antenna beams, the adoption of a reference satellite antenna pattern may be desirable. However, the design of a satellite antenna is influenced by various system parameters, such as size and shape of the coverage area, required minimum gain, limitation of aperture size and flux density, etc. Thus, it is difficult to define a satellite antenna reference pattern applicable to the large variety of complex patterns which may be utilized.

The advantages of frequency reuse may not be fully realized if the control of the satellite beam position is inadequate. However, no substantial reduction in these advantages is likely so long as the spacecraft antenna beam position can be held to within a small fraction (0.2 or less) of its beamwidth. For example, the 0.5° beamwidth achievable today can utilize this technique if the beam position is held to within ±1°, which is feasible.

Satellite antennas should be designed so that the satellite may be repositioned in the geostationary satellite orbit and still provide the required service in its coverage areas.

SATELLITE STATION-KEEPING

When the longitudinal position of geostationary satellites is subject to some uncertainty due to orbital drift or orbital inclination, a reduction in the potential geostationary orbit capacity will result. Capacity is only slightly impaired by moderate orbital inclinations, but is greatly reduced when longitudinal positional drifts approach values comparable with the minimum permissible satellite spacing.

The Radio Regulations require all satellites to be maintained within ±1° of the longitude of their nominal position if this is necessary to prevent unacceptable interference in any other satellite network. They urge that efforts should be made to develop spacecraft and control facilities to achieve a capability of maintaining their positions at least within ±0.5° of the longitude of their nominal position.

REVERSAL OF UPLINK AND DOWNLINK FREQUENCY BANDS

It may be feasible to increase the number of satellites using a pair of frequency bands in a given arc of the geostationary satellite orbit if the frequency assignments are reversed between adjacent satellites, the uplink band assigned for one satellite being the downlink band for the next. This technique may, to some extent, compete with other methods of increasing the capacity of the orbit such as the use of high-gain satellite antennas or polarization discrimination to reduce interference between alternate satellites. It may make necessary some deterioration of the sharing criteria in frequency bands shared with terrestrial services. One source of potential harmful interference is from a satellite located almost on the other side of the Earth, but directly visible by the satellite being interfered with, when reversed frequency bands are used.

If this technique should be found valuable, it would be desirable to determine which shall be the preferred direction of transmission for each frequency band. A responsibility might then be placed upon systems using bands in the mode which is not preferred to be particularly attentive to site screening at earth stations. This would reduce interference to earth stations using the same frequency bands in the preferred mode.

INTERLEAVING RF CHANNEL FREQUENCIES

The extent to which reduction in satellite spacing and improved orbit/spectrum utilization may be achieved by interleaving the carrier frequencies of one satellite with those of a neighboring satellite is critically dependent on the type of modulation (e.g., FM or PSK) and the satellite multiple-access technique (e.g., single carrier or FDMA) applied to the wanted and interfering carriers. The achievable reduction in satellite spacing may be expressed in terms of an improved tolerance to RF interference which, depending on the modulation and satellite multiple-access techniques applied, may vary from about 0 to 12 dB.

For the case of frequency-modulated FDM telephony an improvement in carrier-to-interference ratio is obtained when interleaved carrier frequencies are used. This is of interest in considering the efficiency of use of the orbit. The improvement is found to be up to about 12 dB, depending upon the modulation indices.

For 4-phase PCM/PSK systems, no advantage is generally obtainable by interleaving the carrier frequencies.

When a satellite network employs frequency reuse by polarization or spot-beam discrimination and FDM/FM, frequency interleaving may provide a useful reduction in the discrimination required. However, for networks which do not have frequency reuse, the sacrifice of half of a broadband channel which frequency interleaving requires is objectionable. Furthermore, where a particular system has more than one satellite in operation and where earth stations are required on occasion to operate with one or the other of these satellites, then the application of frequency interleaving between the two satellites would result in increased earth-station complexity and cost. In the space segment, the advantages of a standard satellite design would be lost.

In the case of systems employing a variety of modulation and satellite multiple-access techniques, the maximum interleaving advantage may be achieved only by appropriate coordination and the allocation of traffic or transmission modes to specific satellite RF channels. However, this may not be possible in practice because of the difficulty in accurately forecasting traffic requirements or new applications, and the loss of flexibility in reassigning traffic. As noted, there will be little improvement in satellite spacing requirements to be obtained by interleaving digital signals in such cases, but this is not likely to be a limiting factor, since the spacing required by analog signals is usually greater than that for digital.

In view of the foregoing considerations, the advantages of frequency interleaving between satellites may, in practice, be restricted to relatively few applications.

INTERFERENCE ALLOWANCE

Studies show that capacity improvement can be obtained from the geostationary satellite orbit if a large part of the noise budget is allocated to interference between satellite systems. For example, it is estimated that the total capacity of a busy arc of the orbit might be increased by at least 75% if the inter-system interference noise component were raised from the 10% of total noise recommended for systems using FDM/FM telephony and television to about 50% of total noise.

To increase the inter-system interference noise allowance beyond that recommended in CCIR Recommendations 466 and AC/4 so that it forms a substantial portion of the total channel noise would have serious disadvantages, for example:

1. The capacity of individual satellite networks would be reduced. This would increase costs, especially for satellite systems used to provide international links for the public network.

2. Interference is sensitively dependent upon such parameters as:

    a. accuracy of satellite station-keeping

    b. accuracy of satellite attitude control

    c. sidelobe characteristics of satellite and earth-station antennas

    d. cross-polarization response of antennas.

CCIR studies of most of these parameters are, as yet, in an early stage. If inter-system interference noise is only a small part of the total noise, system designers are able to exercise control over system performance. However, if the interference component were large, it would be unlikely that performance targets would be achieved, as the system designer might have no control.

Eventually, it may be found desirable to increase the recommended inter-system interference noise component beyond the proportion of total noise currently recommended in order to increase the communications capacity made available by the geostationary satellite orbit. This increase might be made selectively, applying in particular to the busier arcs of the orbit. If arcs of the orbit should be occupied by satellites serving earth stations with antennas below minimum specified performance values (which would have to be established) these systems might have to be designed to tolerate high levels of interference noise should close satellite spacing be required.

MODULATION CHARACTERISTICS

For FM systems, as the modulation index is increased, the capacity per satellite is reduced but the baseband noise density due to interference at

a given carrier-to-interference ratio falls, permitting closer satellite spacing and providing an increase in the effective use of the geostationary satellite orbit. For digital transmissions using PSK, similar conditions exist; that is, the interference immunity of a signal is increased as the number of phases is reduced, again allowing close satellite spacing. However, in this case, the utilization of the geostationary satellite orbit tends to be optimized when the number of phases is in the range of four to eight, the orbit utilization tending to decrease as either a higher or a lower number of phases is used.

SYSTEM HOMOGENEITY

The most efficient orbit utilization would be obtained if all satellites utilizing the geostationary satellite orbit, illuminating the same geographical area and using the same frequency bands had the same characteristics. However, in practice, such homogeneity is not available in satellite systems.

Consider two satellite systems, A and B, with satellites in adjacent orbital positions. If A and B have widely differing characteristics (regarding satellite receiver sensitivity and downlink EIRP, or regarding their associated earth-station characteristics), the angular spacing necessary to protect A against interference from B may differ from that necessary to protect B from A. In practice, the greater of the two angles must be selected. The extent to which this may represent an inefficient utilization of the geostationary satellite orbit is dependent on many parameters in the design of the satellite systems using orbital positions near those of A and B. It is possible to improve utilization of the orbit if inhomogeneity is taken into account during the satellite system design. The particular system parameters which should be given consideration are satellite EIRP, earth-station figure of merit (G/T) and the relative immunity of the modulation system to interference.

A review of current system technology gives the following conclusions:

- A reasonably efficient orbit utilization may result if a low EIRP satellite working with earth stations having high-gain antennas were placed between existing satellites of systems with high downlink EIRP and earth stations with low values of G/T.

- With different types of satellite in orbit, efficient orbit utilization might be obtained through a suitable choice of transmission parameters and adjustment of interference noise allowances, relative power levels and orbital positions of the satellites.

- When space networks with significantly different downlink EIRP share the orbit, it is generally advantageous to cluster several low EIRP satellites; the size of such clusters should have some form of inverse relationship to the difference in satellite EIRP.

- The use of spot-beam satellite antennas complicates the problem of orbital capacity.

MULTIPLE FREQUENCY BANDS

Communications satellites use frequency bands in pairs, one band for the uplink and the other for the downlink. Up to the present many systems have paired the use of the 4- and 6-GHz bands because they were allocated early and they provide good propagation conditions. In the future, new frequency bands allocated at the WARC-ST 1971 to various space communications services will be used, and the propagation conditions obtaining on these frequency bands, with the differences in available bandwidth, may lead to preferred pairings of these new bands. It may be desirable to define preferred pairings for frequency bands to avoid wasteful use of the spectrum and the geostationary satellite orbit.

In some satellite networks it may be economically and operationally advantageous to use more than one pair of frequency bands, because this enables the effective bandwidth of the network to be increased and is usually the most economical way of increasing the communication capacity. One pair of frequency bands heavily loaded in the relevant part of the orbit has no significant impact on frequency-spectrum economy or orbit utilization, but multiple banding has certain disadvantages when the second pair of frequency bands is intensively used. These are:

1. The coordination of frequency assignments is increased in complexity and the optimization of the orbital location of satellites operating in the various frequency bands is no longer independent, so that the efficiency of these processes is reduced.

2. Full use of orbit can be made in only one pair of frequency bands because of probable different angular separations required in different pairs of frequency bands.

Significant differences in the angular separation required may arise even from differences in the propagation margin required in different bands, and the situation might be aggravated by differences in such characteristics as type of modulation, transmission parameters or type of antenna. The problem is not significant in parts of the orbit where few satellites use more than one pair of frequency bands, but it could have a substantial adverse effect on orbit/spectrum utilization efficiency where there are several multi-band satellites in adjacent orbital locations. There is insufficient information at this time to quantify fully the effect on orbit/spectrum utilization of using multi-band satellites in the fixed-satellite service.

Two strategies for preventing inefficient orbit/spectrum utilization have been suggested. These are:

1. To minimize the overall orbit/spectrum capacity losses by adjusting system parameters. This generally corresponds to equalizing the required separation angles in the various bands.

2. To make room between two multi-band satellites for an additional satellite operating in only one pair of the frequency bands used on the multi-band satellites. This, however, may involve adjustment of the characteristics and parameters of the satellite networks.

These two strategies should be considered in determining the characteristics and parameters of satellite networks using more than one pair of frequency bands.

Tools For Quantitative Analysis

DONALD JANSKY
Office of Telecommunications Policy
Washington, D.C., U.S.A.

Various techniques interact in complex ways to influence the capacity of satellites and the angular spacing required between them when they use the same frequency bands. No analysis embracing all parameters has been completed yet, but the studies of parameters that have been done are summarized in the following.

ORBIT SPACING STUDIES

Satellite spacing required and the network capacities achievable are calculated for systems with various arbitrarily chosen characteristics, earth-station antennas being assumed to conform to CCIR Recommendation 465. Some results follow (1):

• With earth-station G/T about 40 dB/K, using PSK or wide-deviation FDM/FM, required satellite spacings in a homogeneously occupied arc of the orbit are typically between 1.5° and 3.0°.

• With earth station G/T about 28 dB/K the corresponding satellite spacing required for wide-deviation FM is 3-4°, but only about 1.5° would be required for PSK.

• With low-deviation FM, spacings may be two or three times as great as for wide-deviation FM.

• When a satellite serving earth stations with G/T of 40 dB/K is adjacent to one serving earth stations with G/T of 28 dB/K, the required spacing may be two or three times as great as either network would need in a homogeneously used arc of the orbit.

HOMOGENEOUS PARAMETRIC STUDIES

A homogeneous inter-system interference model has been constructed using the following hypotheses:

• a full ring of equally spaced satellites in the geostationary orbit;

• frequency modulation, with FDM telephony basebands of not less than 240-channel capacity;

• all satellites have earth-coverage antennas, posses identical transmission and modulation parameters, and transmit the same type of signal on the same nominal carrier frequency (co-channel sharing);

• all earth stations are identical, with antenna sidelobe characteristics conforming to Recommendation 465;

• propagation conditions are clear weather;

• the downlink only is considered.

All parameters are allowed to vary freely, and for this reason, the homogeneous model provides direction for the various trade-offs involved.

Figure 2 shows the effect of the interference noise budget N_I on the minimum inter-satellite spacing. The satellite spacing has been normalized to unity at $N_I = 1000$ pWOp, which is the value currently stipulated in CCIR Recommendation 466. Similarly, Figure 3 shows the sensitivity of the minimum inter-satellite spacing to changes in earth-station antenna size. Here, the spacing is normalized to unity at $D/\lambda = 100$, where D is the antenna diameter and λ is the operating wavelength.

The impact of the modulation parameters can be observed in Figure 4. The abscissa is given in terms of n, the number of telephone channels per MHz, which is a function of the modulation index. As in previous curves, the ordinate shows the corresponding change in spacing, relative to the case with an rms modulation index of 1.0. Figure 4 further indicates that a substantial improvement in spacing can result from an interleaved frequency plan. However, for small modulation indices, the co-channel condition is no longer the worst case of a frequency plan. In fact, for relatively small departures from the true co-channel

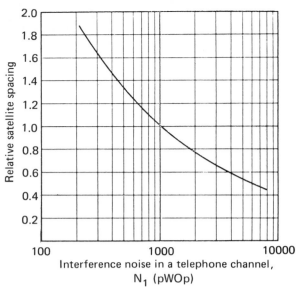

Fig. 2. Satellite Spacing as a Function of
Interference Noise

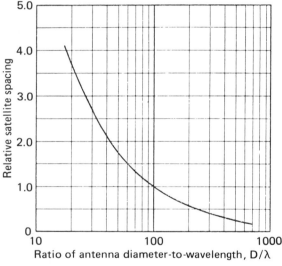

Fig. 3. Satellite Spacing as a Function of
Earth Station Antenna Size

carrier-to-thermal noise ratio (CNR), since it is
more convenient to use than EIRP but is related
to the latter in an obvious fashion. Finally, the
figure shows loci of constant ratio of thermal-to-
interference noise when their sum totals 7500 pW.

Some conclusions may be drawn from Figures
2 to 5. Some of these conclusions, of course,
apply to the FDM/FM homogeneous system
model only, but may be extended to more general contexts.

First, Figure 2 shows that, all other things being equal, satellite spacing varies inversely with
interference noise allowance. Furthermore, the
variation is rapid; for example, as N_I varies from
250 pW to 7500 pW, a 30:1 change, $\Delta\theta$ varies
by about a factor of 4. Thus, as N_I approaches
the total noise budget, there are diminishing re-
turns in $\Delta\theta$ for allowing progressively more interference noise.

Next, Figure 3 also indicates that satellite
spacing varies inversely with earth-station antenna
size, the relative rate of change being fairly small
when antennas are relatively large (D/λ greater
than 100) but more rapid for small antennas.

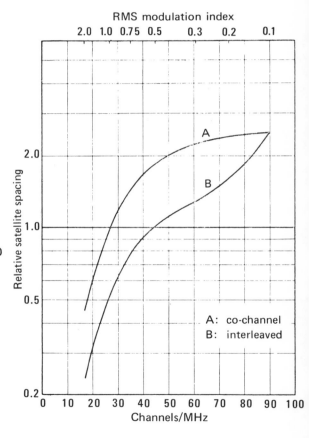

Fig. 4. Satellite Spacing as a Function of
Modulation Index

condition (particularly when the difference between the carrier frequencies is less than the
width of the baseband) the interference can increase markedly.

Figure 2 displays a variation in interference
noise without, however, specifying the corresponding variation in EIRP which must occur to maintain the total noise budget (interference noise
plus thermal noise) at some specified value. An
equitable basis for comparison would seem to require maintaining fixed values of EIRP while
varying the spacing. These interactions are displayed in Figure 5, where ñ, the number of telephone channels per orbit degree and per MHz,
has been used as a meaningful measure of orbit/
spectrum utilization. The parameter chosen is

The relationship plotted in Figure 4 shows
again that, all other things being fixed, satellite
spacing varies inversely with modulation index,

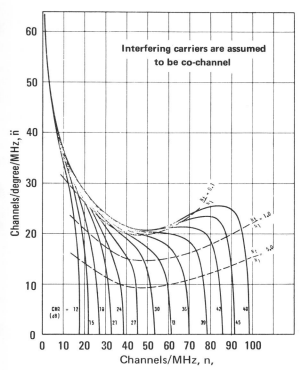

Fig. 5. Orbit/Spectrum Utilization as a Function of Chennels per Satellite (D/λ = 300, $N_t + N_T$ = 7500)

these being roughly in a one-to-one correspondence for a high modulation index. The capacity of each satellite is reduced substantially as the modulation index is increased, but there is a gain in total orbit capacity because more satellites can be used in the orbit. Thus, increasing modulation index is potentially an effective means of increasing orbit utilization in FM/FDM systems.

For the adopted measure of orbit/spectrum utilization, namely channels/degree/MHz, $\ddot{n}$, it can be seen from Figure 5 that efficiency generally increases with modulation index, in agreement with the foregoing observation. However, a high modulation index implies low-capacity satellites for a fixed bandwidth. Hence, for a total traffic capacity more satellites are required, and this may result in an economic burden. Further, for some range of modulation index and relatively high carrier-to-noise ratio, there is relatively little variation in $\ddot{n}$, and therefore, within this range, it is possible to choose satellite capacity as desired without compromising orbit utilization.

Also, as the carrier-to-noise ratio decreases, so the possible range for $\ddot{n}$ decreases. This occurs because for each carrier-to-noise ratio, there is an $\ddot{n}$ for which the thermal noise (N_t) approaches the total noise allowance (N_T), at which point no interference noise is permitted. This, in turn, requires "infinite" spacing. It accounts for the downward bend and asymptotic appearance of the curves. Finally, as expected from Figure 2,

greater total orbital capacity is potentially available if a successively larger fraction of the noise budget is allocated to interference.

THE EFFECTS OF STATION-KEEPING ACCURACY ON ORBITAL CAPACITY

As the various methods available are used to reduce the minimum necessary spacing between satellites, the impact of errors in station-keeping becomes increasingly serious. A study of homogeneous orbit occupation by FDM/FM transmission systems has been made, some of the results of which are shown in Figure 6. The curves show that large improvement in total orbit capacity may be obtained both by increasing the interference noise allowance and by increasing the modulation index, assuming that the satellites remain on their assigned positions within about ±0.2°. However, not only is the capacity of the orbit reduced, with transmission parameters unchanged, but the advantage of increasing the interference noise allowance is greatly diminished if satellites drift as far as 0.5° from their nominal positions. The trend toward a higher modulation index may even be reversed.

OPTIMIZATION OF HETEROGENEOUS SATELLITE SYSTEMS

In general, the satellite networks operating in the same band and making use of adjacent positions on the geostationary satellite orbit have dissimilar characteristics. This situation exists in fixed-satellite bands, and where the fixed-satellite bands are shared with other services, such as the Broadcasting Satellite Service. It seems probable that this situation will not prove to be adaptable to analytical methods which apply to arbitrary groupings of satellites. Such situations seem best suited to case-by-case solutions, whereby different combinations of satellites are analyzed to determine the optimum arrangement, and hence are amenable to treatment by computer.

Optimum arrangement analysis was applied using some of the preliminary characteristics of several proposed U.S. domestic satellite systems. Certain characteristics of the Canadian domestic system also were considered, although the characteristics assumed do not take into account all present operational parameters of the Canadian domestic system (including the critical FDMA case). The stated assumptions demonstrated that, despite appropriate positioning of satellites, such a heterogeneous set of satellites cannot achieve acceptable interference levels at a uniform spacing of 3°, even with polarization discrimination. However, these same systems could use an average spacing of about 3° with careful coordination. To the extent that the actual system parameters differ from those assumed, the results would have to be reexamined on a case-by-case basis. An ideal situation would exist in

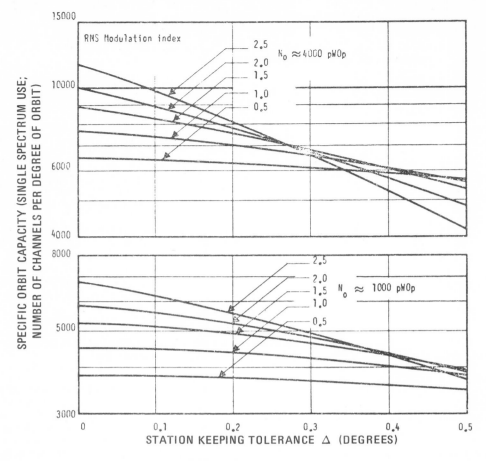

Fig. 6. Specific Orbit Capacity vs Station Keeping Tolerance for Two Values of Interference Noise
Allowance, with RMS Modulation Index as a Parameter

- Homogeneous system
- Global satellite antenna beams
- Earth station antenna diameter 26 m (85 feet)
- CCIR earth station reference antenna pattern
- Bandwidth = 500 MHz

a homogeneous set of FDM/FM networks, having earth-station antennas not less than 10 m (30 feet) in diameter, which could achieve 2-3° orbital spacing.

Analytical methods such as those described provide an effective means for optimizing the positioning of satellites in networks having dissimilar characteristics in congested areas of the geostationary satellite orbit, whether of the same or different services. Of course the orbital positions can most easily be optimized when all systems are in the planning state.

References

1. Orbit-Spectrum Utilization Studies, General Electric Space Systems Organization 1969-1970 (PB 194, 780, 781, 782, 203, 388, 389).

World Administrative Radio Conference For Space Telecommunications (WARC-ST) 1971

J. D. PALMER
Department of Communications
Ottawa, Canada

RESULTS OF THE WORLD ADMINISTRATIVE RADIO CONFERENCE

The World Administrative Radio Conference (WARC), typical of those held about every eight or nine years, met in 1971 in Geneva, Switzerland. Representatives from 101 countries, all members of the International Telecommunication Union (ITU), attended. The primary purpose of this meeting, also known as WARC-ST 1971, was to revise the international Table of Frequency Allocations to accommodate new space services, and to establish or update the conditions of use for space services.

CHANGES TO THE TABLE OF FREQUENCY ALLOCATIONS

New frequency bands were allocated to various satellite services. A summary of the frequency allocations as they now stand for the Fixed Satellite and Broadcasting Satellite Services in Region 2 (the geographic area consisting of North America, South America and Greenland) is given in Table 1. The word "up" indicates that the frequency is allocated to the ground-to-satellite link, and the word "down" indicates that the frequency is allocated to the satellite-to-ground link. Table 2 indicates the allocations for other space services, and Table 3 lists those for space operations and inter-satellite working.

Tables 2 and 3 represent only a *summary* of available bands for each service, and the reader is referred to the Final Acts of the World Administrative Radio Conference for detailed information on the conditions under which each of these frequency bands may be used.

CHANGES IN CONDITIONS OF USE
(Power flux-density limits)

Included in the revised conditions of use are the new power flux-density limits for satellites in various space services in the shared frequency bands. The previous Conference in 1963 (EARC) had established flux limits for space stations in the Communication Satellite Service (now called the Fixed-Satellite Service), for both total flux density and flux density in any 4-kHz band. This conference deleted the limitation on total flux and established limits for flux density in given specified bandwidths. These new limits depend on both frequency and angle of arrival. They are shown in Table 4.

EARTH STATION EIRP

Also changed were limits for earth-station EIRP. In the band 1-15 GHz, EIRP in the horizontal plane radiated by an earth station is limited to:

40 dBW in any 4-kHz band, if the actual horizon is below the horizontal plane;

$40 + 3S$ dBW in any 4-kHz band, where S is the angle of elevation above the horizontal plane between 0 and $5°$; above 15 GHz, the values become 64 dBW in any 1-MHz band for horizon below horizontal plane, and $64 + 3S$ dBW in any 1-MHz band for horizon above horizontal plane.

For angles of elevation greater than $5°$, there is no restriction on EIRP at any frequency.

TABLE 1. Frequency Allocations for the Fixed Satellite and Broadcasting Satellite Services in Region 2.

Fixed Satellite	Frequency	Broadcasting Satellite*
	620–790 MHz	down
down	2500–2535 MHz	down ⎫
	2535–2550 MHz	down ⎬ domestic & community
	2550–2655 MHz	down ⎭
up	2655–2690 MHz	down
down	3400–3500 MHz	
down	3500–3700 MHz	
down	3700–4200 MHz	
up	4400–4700 MHz	
up	5925–6425 MHz	
down	6625–7125 MHz	
(Brazil, Canada, US only)		
down	7250–7300 MHz	
down	7300–7450 MHz	
down	7450–7750 MHz	
up	7900–7975 MHz	
up	7975–8025 MHz	
up	8025–8400 MHz	
down	10.95–11.2 GHz	
down	11.45–11.7 GHz	
down	11.7 –12.2 GHz	down
up	12.5 –12.75 GHz	
up	14.0 –14.3 GHz	
up	14.3 –14.4 GHz	
up	14.4 –14.5 GHz	
down	17.7 –19.7 GHz	
down	19.7 –21.2 GHz	
up	27.5 –29.5 GHz	
up	29.5 –31 GHz	
down	40–41 GHz	
	41–43 GHz	down
up	50–51 GHz	
	84–86 GHz	down
up	92–95 GHz	
down	102–105 GHz	
up	140–142 GHz	
down	150–152 GHz	
	220–230 GHz	
	265–275 GHz	

*Transmissions to satellites in this service fall under the Fixed Satellite Service.

CHANGES IN THE METHOD FOR CALCULATING COORDINATION DISTANCE

The new frequency coordination procedures adopted by the WARC reflect the experience gained in this area since the 1963 EARC. While definitive methods for assessing the likelihood of interference between stations operating in shared frequency bands were not established, the WARC did adopt the generalized method for generating coordination contours promulgated by the Special Joint Meeting (SJM) of the CCIR in February, 1971. This new method is applicable to the frequency range from 1 to 40 GHz, whereas the previous range was 1 to 10 GHz. In addition, the possibility of interference due to precipitation scatter mechanisms has been taken into account.

In this new method, which is applicable to all space and terrestrial services operating in shared frequency bands, the maximum permissible interference power is calculated in terms of the thermal noise power in the reference bandwidth of the receiving system sustaining interference, rather than in terms of the desired signal power (the method given by the EARC). In practice, the methods of the EARC were updated in the intervening years. CCIR Report 382, for example, used thermal-noise power as the criterion for interference to receivers in terrestrial microwave systems, while retaining the use of desired-signal power levels in the case of interference from terrestrial transmitters to receivers at satellite earth stations. Although this new method gives values for the maximum permissible interference power that differ little from

TABLE 2. Mobile-Satellite Frequency Allocations

Frequency	Service		
	Aeronautical Mobile-Satellite	Maritime Mobile-Satellite	Mobile-Satellite
406–406.1 MHz			E
1535–1542.5 MHz		E	
1542.5–1543.5 MHz	S	S	
1543.5–1558.5 MHz	E		
1636.5–1644 MHz		E	
1644–1645 MHz	S	S	
1645–1660 MHz	E		
43–48 GHz	S	S	
66–71 GHz	S	S	
95–101 GHz	S	S	
142–150 GHz	S	S	
190–200 GHz	S	S	
250–265 GHz	S	S	

S = shared   E = exclusive

TABLE 3. Space Operations and Inter-Satellite Frequency Allocations

Frequency	Inter-Satellite Working	Telecommand	Satellite Identification	Telemetry & Tracking
30.005–30.010 MHz			S	
137–138 MHz				S
267–273 MHz				S
400.15–401 MHz				S
401–402 MHz				S
1427–1429 MHz		S		
1525–1535 MHz				S
54.25–58.2 GHz	E			
59–64 GHz	E			
105–130 GHz	E			
170–182 GHz	E			
185–190 GHz	E			

S = shared   E = exclusive

previous values, this new method frees the value of the maximum permissible interference power from dependence on the parameters of the transmitter in the link being interfered with. Perhaps the most noticeable difference between the EARC and the WARC method lies in the form and content of the basic propagation data.

In the 1963 EARC method, transmission loss was expressed as the sum of two independent terms: a basic transmission loss which is a function of distance, and a site shielding factor which is a function of the horizon elevation at the earth-station site. These terms are now combined in a single, basic transmission-loss term expressed as a function of both distance and horizon elevation angle. Propagation curves adopted at the 1963 EARC were based on essentially worst-case empirical data extracted from measurements made by several administrations. The curves of the SJM (for example, those in Annex 10-1 of the Final Report) were based on the model path geometry shown in Figure 7. An example of transmission losses calculated by both EARC and WARC methods is shown in Figure 8. Note that in the WARC curves, the range of distances has been extended down to 25 km (15 miles).

The new SJM propagation data contained in Appendix 28 of the WARC Final Acts form the basis for the determination of the coordination area around an earth station.

These SJM data are in the form of two curves for each of the three radio-climatic zones, one showing basic transmission loss exceeded more than 20% of the time, and the other showing loss exceeded for more than 0.01% of the time. Loss is given as a function of distance with horizon elevation angle as a parameter. While such curves can be used directly for the calculation of transmission loss and coordination distance by computer, the correction of these curves for atmos-

TABLE 4. Power Flux Density Limits

Frequency Range	Power Flux Density (dBW/m^2)			Reference Bandwidth	Remarks
	$0 \leqslant \theta \leqslant 5°$	$5° \leqslant \theta \leqslant 25°$	$25° \leqslant \theta \leqslant 90°$		
1670–2535 MHz	–154	$-154 + \left(\dfrac{\theta - 5}{2}\right)$	–144	4 kHz	
1690–1700 MHz	–133	–133	–133	1.5 MHz	For Met-Sats sharing with Met Aids
2500–2690 MHz	–152	$-152 + 3\left(\dfrac{\theta - 5}{4}\right)$	–137	4 kHz	Applied to broadcasting satellites in bands shared w/fixed or mobile
3400–7750 MHz	–152	$-152 + \left(\dfrac{\theta - 5}{2}\right)$	–142	4 kHz	
8025 MHz–11.7 GHz	–150	$-150 + \left(\dfrac{\theta - 5}{2}\right)$	–140	4 kHz	
11.7–12.75 GHz	–148	$-148 + \left(\dfrac{\theta - 5}{2}\right)$	–138	4 kHz	
17.7–22.0 GHz	–115	$-115 + \left(\dfrac{\theta - 5}{2}\right)$	–105	1 MHz	

θ = angle of arrival (°) above the horizontal plane

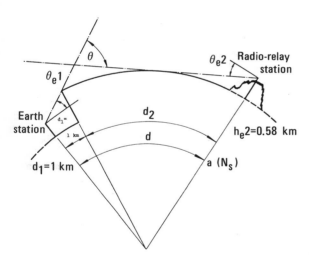

Fig. 7. Model Path Geometry (Taken from Annex 10-1, Final Report, SJM, CCIR, FEB. 1971)

1. unit elevation-angle correction as a function of required normalized coordination loss, with frequency as a parameter;

2. total elevation-angle correction as a function of the unit elevation-angle correction, with horizon elevation angle as a parameter; and

3. coordination distance as a function of frequency, with required corrected coordination loss as a parameter.

Another significant change in the coordination procedure embodied in the Final Acts of the WARC is the inclusion of the effects of precipitation scatter via main-beam, common-volume, inter-

pheric absorption at frequencies above 10 GHz makes it difficult to use them in graphical calculations, the atmospheric absorption factor being a complex function of frequency and distance. To simplify the presentation of these propagation effects, the WARC adopted the form shown in Chapter 8 of the Final Report of the SJM; that is, a set of three curves for each radio-climatic zone. These appear as Figures 4 to 12 of Appendix 28 of the WARC Final Acts. The three curves given for each zone are:

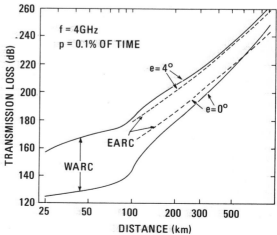

Fig. 8. Basic Transmission Loss vs Distance

sections which may be greater, for small percentages of time, than interference propagated by the tropospheric scatter mode.

The EARC coordination procedure was based on the assumption that non-geostationary satellites would be employed. The WARC methods take geostationary satellites into account in several ways. For example, since earth stations direct their main beams over a limited range of azimuth and elevation angles when working with geostationary satellites, advantage can be taken of the lower antenna gain in all other directions. Also, the percentage of time for which the calculation should be made for high levels of interference was reduced from 0.1% to 0.01%.

The EARC considered only the range from 1 to 10 GHz, and treated three cases of interference: from a terrestrial station to a communications-satellite earth station; to a space-research earth station; and from a communications-satellite earth station to a terrestrial station. The WARC methods, on the other hand, apply to many more possible interference situations. Typical system characteristics are given in Tables I and II of Appendix 28.

Lastly, the WARC coordination procedure takes into account atmospheric absorption, a phenomenon which becomes increasingly important as the frequency increases.

Part 4

Earth Station Technology

Earth Station Technology

E. R. WALTHALL
RCA
Princeton, N.J., U.S.A.

INTRODUCTION

Since the advent of the first experimental communications satellites TELSTAR and RELAY launched in 1962, more than a decade of communication satellite system development has passed with the development of the geostationary satellites typified by the Intelsat communications satellites. During this period both the space segments and earth station technology have evolved from a performance, reliability and cost standpoint. This evolution has contributed to the expanding use of satellites for communications on a worldwide basis. The most recent event in this expanding use has been the exploitation of domestic communications satellites for interstate telecommunications and television distribution. The basic economy of providing these functions by satellite systems has been demonstrated especially for areas of the world where there is a requirement to reach small, widely dispersed settlements with few transportation facilities and frequently over terrain whose features impede normal construction techniques (1,2).

Thus, the earth-station technology has developed to encompass not only the high-performance, high-capacity stations of the Intelsat network but smaller, much less expensive ground stations designed for totally unattended operation (3). While the evolution of the Intelsat system has led to the large increase in the number of earth stations in that system (approximately 80) the major expansion in terms of numbers of earth stations for the next decade will be dominated by the regional and domestic systems providing both point-to-point communications and direct broadcast for television distribution (4,5). As a result, a major impetus for the development of earth-station technology will be provided by the need for low-cost earth terminals. Mass-production techniques of simple designs employing small non-tracking antennas, uncooled parametric amplifiers, transistorized receivers, maximum use of integrated-circuit technology and large scale integration (LSI) techniques, efficient modulation schemes and demand-assignment network switching will typify the earth stations, particularly for domestic and regional systems (6).

Table 1 lists some of the key features that presently characterize earth stations ranging from the large Intelsat stations, to medium-sized stations for use with domestic systems such as the Canadian Telesat system, to minimum-sized stations that might be used for thin-route telecommunications. Table 2 lists the characteristics of a typical direct-broadcast receive-only earth station being developed for the Federation of Rocky Mountain States, Inc. (7). These stations are being used for an educational TV experiment in conjunction with the ATS-6 satellite.

EARTH-STATION SYSTEMS PLANNING

The major requirement for the satellite telecommunication system is that it provides reliable, telephone-quality, circuits to remote areas. Earth stations must be rugged and easy to install in remote areas. Since there is usually a large number of earth stations, unit station cost has a great impact on total system cost, and considerable emphasis is placed on the design of low-cost stations. Thus, in developing a system approach, tradeoffs are made between the ground-station performance and cost for a given satellite capability; i.e., the ground-station figure of merit, station capacity and link quality is optimized for a given satellite EIRP and bandwidth availability (8,9,19). Typically, the satellite antenna gain is restricted by the desired area coverage and

TABLE 1. Earth Station Characteristics

High Performance Stations	Medium Performance Stations	Minimum Cost Stations
Antenna • Wheel and track azimuth control • Single elevation bull gear • No upper equipment room • No access elevator • All electronics in base of elevated foundation Tracking System • Step-track (or "hill climber") tracking system instead of 3-channel monopulse Feed System • Feed horn and feed microwave circuitry at base of antenna • 3-reflector focussed-beam system illuminates sub-reflector through aperture of main reflector Low Noise Receiver • 50 K uncooled paramp mounted at antenna base High Power Amplifier • Tuned klystron or wide-band TWT GCE System • Micro-integrated circuitry • FDM/FM, PCM/TDM Redundancy • All critical components employ on-line redundancy	Antenna • Fixed position • Manual steering adjustment in elevation and azimuth • Az-el or polar mounts Feed • Sum mode feed Low Noise Receiver • 100 K uncooled paramp High Power Amplifier • Wide-band TWT GCE • For medium to high capacity trunking-FDM/FM techniques • For Thin-Route operations - FM compandored or digital single-voice-channel per carrier technique Power System • Battery bank/static inverter backup to commercial power or diesel-generator Redundancy • Redundant LNR and HPA	Antenna • Single element-spun dish fabrication • Fixed position manually steered Feed • Primary focus feed Low Noise Receiver • 170 to 300 K uncooled paramp High Power Amplifier • TWT GCE • FM compandored or digital single-voice channel per carrier technique Power System • Battery bank/static inverter backup to commercial power or generator Redundancy • No redundancy with minimum spare parts

TABLE 2. Receive-Only Terminal Characteristics (ATS-6 HEW Experiment)

Frequency band:	2.5/2.6 GHz
Antenna:	3-m (10-foot) diameter (segmented assembly)
Antenna tracking*:	Operator controlled electric drive elevation track $\pm 3°$
Antenna feed:	Prime focus feed with built-in preamplifier (circular polarization)
Receiver:	Hybrid microwave integrated circuit tuned-radio frequency receiver with 4.2 dB noise figure
Figure-of-merit:	7.1 dB/K

*Tracking is required since north-south stationkeeping is not employed by the ATS-6 spacecraft.

the maximum transmitter power capability is limited by the saturation level of the RF power amplifier and the availability of power from the satellite power supply. Furthermore, the CCIR-allowable flux densities for the allocated frequency bands limit the maximum satellite EIRP to values that do not produce interference with terrestrial microwave facilities (10).

The minimum antenna diameter (and consequently the minimum antenna gain) are determined by the requirement for spatial isolation of satellites in close proximity on orbit and by the desire for non-tracking antennas for thin-route operations and remote sites. The availability of receiver front-end noise temperatures ranging from 20 K for cooled parametric amplifiers (used in the high-capacity Intelsat stations) to the 50 to 300 K range uncooled parametric amplifiers (used in the smaller and less costly domestic-satellite earth terminals) offers a range of station cost and performance capabilities that varies by more than two orders of magnitude (8,9,11,12,20). In some cases transistorized front ends with a noise figure of 3 to 4 dB have been used to reduce cost (7). Thus, commercially planned satellite earth-station figures of merit range from 18 dB/K to 41 dB/K.

Modulation techniques and access methods such as those given in reference (13) and shown in Table 3 may be selected to optimize channel capacity and system cost based on the satellite and ground-station performance capabilities. Figure 1, taken from reference 14, compares several modulation techniques with various ground-station figures of merit for a thin-route telecommunications service. While the data presented are based on a particular satellite and ground-system configuration, similar tradeoffs can be conducted to optimize any particular set of telecommunications requirements.

Computer programs have been developed which analyze and evaluate the major system parameters of a domestic satellite communications system such as type and quality of service, earth-station figure of merit, satellite EIRP, and component costs (17). These programs combined with other programs that analyze the mix between terrestrial facilities and satellite facilities provide a means for optimizing the overall space-system design. Reference (18) provides a description of the telecommuni-

TABLE 3. Modulation and Access Methods

Access Method	Modulation Technique	Channels Per Transponder B_{IF} = 36 MHz	RF Power Per Transponder	Availability	Comments
FDMA	AM SSB SC SCC	7200	$\sim$ 150 W	10 Years	Requires NPR of 42 dB for tolerable intermodulation
	AM DSB SC SCC	3600	$\sim$ 75 W	10 Years	Same as above
	FM SCC	360	$\sim$ 10 W	2 Years	VOX required to reduce power and intermodulation
	PCM SCC 4 ϕ PSK	800	$\sim$ 5 W	1 Year	VOX required, 7 bit PCM with synchronous 64 kb/s
	VSDM SCC 4 ϕ PSK	1200	$\sim$ 5 W	1 Year	VOX required. VSDM at 40 kb/s
	VSDM SCC DC PSK	600	$\sim$ 5 W	1 Year	VOX required. VSDM at 40 kb/s
TDMA	PCM 4 ϕ PSK	800 892	$\sim$10 W	4 Years	Limited to 50 M b/s by logic circuits. 72 M b/s possible with 4 ϕ PSK
	VSDM 4 ϕ PSK	1500 1784	$\sim$ 10 W	4 Years	Same as above

General Conditions: U.S. Continental Coverage, 10-m (33-ft) Antenna and Uncooled Paramp in Earth Station Demand Assignment is fully-variable for all access methods

SYMBOLS

FDM	Frequency Division Multiplex	PCM	Pulse Code Modulation
TDM	Time Division Multiplex	DC	Differentially Coherent
AM	Amplitude Modulation	PSK	Phase Shift Keying
SSB	Single Side Band	VSDM	Variable Slope Delta Modulation
SC	Suppressed Carrier	B_{IF}	If Bandwidth
SCC	Single Channel Per Carrier	RF	Radio Frequency
FM	Frequency Modulation	VOX	Voice Activated Carrier
ϕ	Phase		

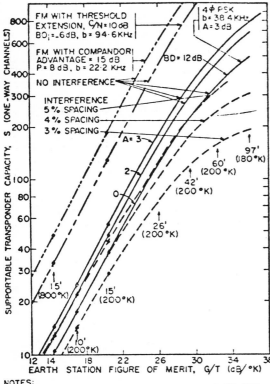

NOTES:
EXCEPT WHERE STATED OTHERWISE ALL CURVES ARE
FOR BO_i = INPUT BACK-OFF = -10 dB
b= RF BANDWIDTH = 76.3 KHz
FREQUENCY MODULATION VOICE CHANNEL QUALITY =
37.5 dBrnCo
P = PRE-EMPHASIS = 4dB PEAK VOICE SIGNAL = 3dBmO
NO COMPANDOR ADVANTAGE
A = INTERMOD ADVANTAGE = 3 dB

SINGLE CARRIER SATURATED VALUES:
UP-PATH C/N_o = 104.6 dB-Hz
DOWN-PATH C/N_o = 101.1 dB-Hz $(G/T = 36)$

**Fig. 1. Supportable Transponder Capacity Versus
Earth Station G/T**

cations requirements for Brazil and the planning
for a satellite system that is presently being
considered for a major expansion of its existing
network.

REVIEW OF EARTH STATION DESIGN AND DEVELOPMENT

Figure 2 is a block diagram of a typical ground
station. The major hardware subsystems can be
listed as follows:

Antenna
Antenna feed
Low-noise receiver (LNR)
High-power amplifiers (HPA)
Ground communication equipment
Interface equipment
Monitoring and control equipment

A description of a recently designed Intelsat station can be found in reference 15. Descriptions of
ground stations related to domestic satellite applications and small earth terminals for direct broadcast reception may be found in references 3, 5, 6,
7, 11, 12, 19 and 20.

Evolution of Commercial Satellite Earth Stations for Operations with the Intelsat Global Network

The complement of personnel required for
operation and maintenance of Intelsat-type commercial satellite earth stations has been greatly
reduced with the new terminals now available in
comparison to the pioneer stations of the early
1960's; for example, the initial installations
at Goonhilly Downs, Cornwall, England, Andover,
Maine, Pleumeur Bodou, Brittany, France, Raisting
(near Munich, FRG) and Mill Village, Nova Scotia,
Canada. For these pioneer stations the average
complement of personnel for operation and maintenance was about 35. Some new terminals
require a staff of about 15; some require a staff of
less than 10, and the most recent are unattended,
requiring only periodic visits for routine maintenance work.

Over the last 8 years improvements in design of
earth stations have permitted over a 50% reduction
in the operating and maintenance staff, and have
reduced the cost of the earth stations by more
than 50%.

More than 80 Intelsat earth stations in the
developing nations stimulated the demand for low-
cost earth stations that are easy and economical
to operate and maintain. This need has been ful-
filled by competition from earth-station contractors from Great Britain, Canada, Italy, Japan and
the United States.

Earth-station design has evolved in steps to the
present level of capital costs and operating and
maintenance costs. Four design phases have been
selected to describe the rapid evolutionary process
in earth-station development. The design phases
necessarily overlap to a certain degree.

Phase 1 Earth Stations

Phase 1 was represented by the early Intelsat
stations in Great Britain, U.S., France, Germany
and Canada, some of which participated in the Telstar and Relay Satellite experiments of 1962 and
1963 that demonstrated the feasibility of a global
satellite communication system. These stations
started commercial trans-Atlantic satellite service
with INTELSAT I (Early Bird), the first Intelsat
satellite. All these stations, except Goonhilly,
Great Britain, had temperature-controlled radome
protection to the antenna to assure tracking accuracy in the presence of wind gusts, icing and wide
variations in ambient temperature, all of which
would bring a degree of deformation to an exposed
reflector. The radomes, made of hyperlon-coated
dacron, were air inflated at slightly above atmospheric pressure. These stations also had a control
building for the station separated from the antenna
complex by about 480 meters (1500 feet). This
separation was designed for the possibility of track-

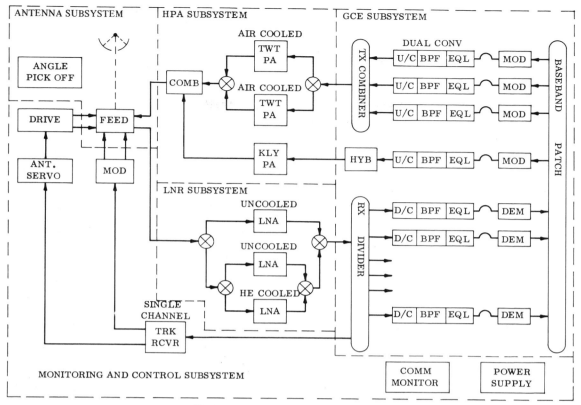

Fig. 2. Typical Ground Station Block Diagram

ing sub-synchronous satellites where two or more antennas, with a centrally located control building, would not shadow another antenna at low radiation angles. Receiving capacity was restricted to only a few RF carriers and the interfacility link was at the baseband frequency level, from RF and IF signal processing equipment at the antenna location to control consoles in the control building. Antenna reflector sizes varied from 26 m (85 feet) to 30 m (100 feet). The high-power amplifier used liquid-cooled 8 kW klystrons—a power level necessary for operation with INTELSAT I. The low-noise receiver subsystems for these early stations were narrow-band masers except for Mill Village No. 1 in Canada which used a liquid-helium parametric amplifier.

Some of the Phase 1 stations had a cone type of upper equipment room for feed, tracking downconverter and low-noise receiver which moved in elevation and azimuth. This made access and maintenance of these subsystems particularly awkward.

Phase 2 Earth Stations

Phase 2 stations dispensed with a radome-protected antenna by improvements in the antenna structure and tracking system, and by the ability to obtain the exact degree of antenna reflector deformation caused by winds, ice and temperature extremes. Electric de-icing systems consisting of thermostatically controlled electric blankets behind the reflector panels were used in locations where ice and snow were of concern. During this phase the interfacility link between the antenna complex and the control building became a continuous, flexible, elliptical waveguide system carried on a trough a few feet above ground level—one waveguide for the 6-GHz transmit signals and one for the 4-GHz receive carriers. All signal processing was carried out at the control building, from baseband to RF in the transmit path, and RF to baseband in the receive path. The 4-GHz waveguide could carry over 20 receive carriers, or one from each station to which communications were required. This reconfiguration greatly reduced the operation and maintenance effort in that the RF and baseband signal processing equipment (up and downconverters, modulators, demodulators, etc.) were rack mounted directly adjacent to the control console and in view of the operating personnel, in the control building. Visits to the remote antenna complex were reduced to periodic maintenance checks of the servo-drive subsystem, the low-noise receiver and the high-power amplifier.

Phase 2 stations generally used 3-kW air-cooled klystrons or travelling-wave tubes which represented a significant reduction in maintenance from the liquid-cooled versions. All stations used a helium parametric amplifier with 500-MHz instantaneous bandwidth, 30 dB gain, and 17 to 20

125

K noise temperature in the low-noise receiver. This revision brought a considerable reduction in station maintenance and improvement in station reliability. Many Phase 2 stations used the double-conversion principle in IF/RF signal processing to provide 500-MHz bandwidth and to enable rapid frequency change by crystal selection alone without replacing or retuning filters. Additionally, the double-conversion principle reduced in-band spurious signal generation to an undetectable level. (Spurious signals can reduce the effectiveness of a single-conversion system, particularly if more than about five carriers are present.) The double-conversion scheme in the up- and down-converters permits instead a 1-for-10 redundancy level in receive chains of the 1-for-1 inherent in single conversion. Logic circuitry permits automatic insertion of the standby chain, with frequency and capacity level automatically selected in place of any faulty chain. Improved access to the low-noise receiver and tracking downconverter mounted behind the apex of the antenna was necessary for Phase 2 stations with an exposed antenna, particularly for northern locations where rain, snow and ice would make staircase access hazardous. Accordingly, most of these stations were fitted with an elevator (in the king post of the azimuth mount) from ground level to an upper equipment room that moved only in the azimuth plane. In this room the parametric amplifier and the feed moved in an elevation arc only and remained accessible to maintenance personnel. Phase 2 stations were represented by second stations at Goonhilly, Raisting, Pleumeur Bodou, Fuccino (Italy), and Mill Village; Comsat stations at Paumalu, Hawaii, Brewster Flat, Washington, Etam, West Virginia, and Jamesburg California. Other Phase 2 stations were built in Chile, Hong Kong, Bahrain, Kenya, Brazil, Argentina, Malaysia, Singapore, Australia, New Zealand, India, Pakistan, Morocco, Greece, Spain, Scandinavia, Kuwait and Israel.

Phase 3 Earth Stations

Phase 3 earth stations appeared in about 1968, primarily in the developing areas of the world where the communications traffic capability was restricted to approximately six receive carriers. The king-post configuration of the antenna pedestal was replaced with a wheel-and-track configuration for all Phase 3 stations as an economy measure, although many Phase 2 stations used the wheel-and-track configuration. The main difference between Phase 3 and Phase 2 was the integration of all equipment and station control functions, as well as administration, within the antenna complex. This was provided by an extension of the antenna foundation from ground level to a height of about 4 meters (12 feet) to provide a circular room of 167 to 232 m^2 (1800 to 2500 ft^2) with the azimuth circular-rail system mounted on the roof of the foundation structure. The basement-type room at the base of the antenna was finished with panelling, walls and partitions

to provide a pleasing environment for administration and operation of the country's satellite terminal facility. The concentration of all station personnel at a single location resulted in cost benefits for administration and operation and maintenance of the station. The only active electronic elements not in immediate view of the operating personnel were the low-noise receiver, the tracking downconverter and the high-power amplifier, which were mounted in equipment rooms behind the apex of the antenna and adjacent to the feed system. Some Phase 3 stations, however, had the high-power amplifier mounted at the antenna base and readily accessible by an elevator. Additional space for repairs, offices, stores, conference room, kitchen and sleeping quarters was provided as required by a simple structure extending outward from the periphery of the circular main room.

Some Phase 3 stations had the azimuth rails at ground level and the equipment and operating rooms mounted above the wheels between the legs of the alidade structure. This configuration reduced the amount of concrete work at the site and permitted a prefabricated design which helped to shorten the installation schedule. Furthermore, such a building lends itself to modular design so that equipment can be installed and tested in the modules at the factory for direct shipment to the site. Modular construction reduces cabling, wiring and testing work at the site, and also reduces shipping costs.

The emergency diesel power plant was contained in a separate building adjacent to the antenna base.

Reliability was improved in the Phase 3 stations and spare parts and maintenance costs reduced by replacing the pump and line-amplifier TWT's in the front end. These are replaced with a solid-state pump source for the low-noise receiver and transistorized line amplifiers with 40-dB gain.

Phase 3 type earth stations are typified by those in Iran, Lebanon, Alaska, Guam, Thailand and Philippines.

Phase 4 Earth Stations

The latest generation of earth stations, (Phase 4) concentrated all of the electronic equipment at the antenna base. The Japanese earth-station contractors made this configuration popular, although Raisting No. 2 was the first to use an offset focussed-beam feed system with an elliptical reflector. In Phase 4 stations the radiating source of the feed subsystem is mounted at the antenna base, as is the low-noise receiver, high-power amplifier, and tracking downconverter. The RF connection to the feedhorn is made by either a 4-reflector focussed-beam system contained in a shroud, as used by the Japanese, or by an oversize-waveguide system with transition elements at either end, as used by RCA. This concentration of all electronic/electric equipment at the antenna base eliminates the upper equipment rooms with their associated lighting and air conditioning, the access elevator, remote-control units for the high-power amplifier and low-noise receiver subsystems, and control, power and signal cabling in the antenna tower.

The 4-reflector focussed-beam system has had limited service to date. Problems were experienced in maintaining identical beam axes for the 4- and 6-GHz antennas. With the oversize waveguide approach, only the feed horn is mounted at the apex of the main reflector, the radiating source being mounted at ground level along with the low-noise receiver and high-power amplifier subsystems.

Some configurations of a Phase 4 station use a square building approximately 20 m (64 feet) on a site instead of a circular structure, to enhance the interior appearance and provide increased space for all station functions.

Stations completed by a Japanese contractor in Nicaragua and Ecuador are typical Phase 4 stations. Some new versions of Phase 4 stations utilize an on-off step auto-tracking system instead of the earlier monopulse autotrack system which operated continuously. Step-track locks the antenna to the satellite by maxima seeking of the beacon or communications signal. The step-track system used on the heavy-route Telesat Canada stations and proved in practice at Comsat's Paumalu station, eliminates the tracking downconverter, the 3-channel tracking receiver and the difference-mode circuitry of the feed system. It simplifies the servo-drive system and extends the lifetime of the drive by the on-off mode of operation. Thus, maintenance of the station is simplified considerably through elimination of the units no longer required by the incorporation of step-track. The simplification is of such an extent as to permit operation of the earth station from a remote control center. This type of operation was pioneered by stations in Nicaragua and Ecuador.

With automatic switchover to redundant standby units in the case of the low-noise receiver, high-power amplifier and other units in the transmit and receive path, periodic maintenance is practical. The time and type of failure of the main unit at the time of transfer to the standby unit is shown at the remote-control center. Remote operation of low-capacity earth stations is improved also by simplification of the high-power amplifier. Power levels of only 30 W to 300 W per channel are required for limited message transmission. This enables the use of 500-MHz broadband TWT's without the necessity for retuning when new transmit-frequency assignments are made.

The helium-cooled parametric amplifier remains as the one element in the earth station that is particularly sensitive to maintenance, although with solid-state pump sources and transistorized driver amplifiers this subsystem is achieving high reliability. However, the ground communications equipment, the multiplex and the rearward microwave link are fully solid state and have a high level of reliability. Some new stations are attaining simplification and cost reduction in the low-noise receiver subsystem by having an uncooled standby unit with 50 K noise temperature as back-up to a cooled main unit with 20 K noise temperature that operates with a reflector about 31 m (95 feet) in diameter. Due to the increased EIRP levels of the latest satellite (INTELSAT IV), systems now have over 3 dB more weather margin than required and the necessity for a high G/T has been diminished. This effect is utilized for stations with satellite look angles of 25° or more, where an uncooled parametric amplifier is adequate and imposes no degradation in receive signal level. The 50 K paramp is achieved with a new order of high pump frequencies and indicates eventual full adoption of uncooled units.

Earth stations that have been installed recently for high-density traffic in the developed nations are designed with particular emphasis on high reliability. Consequently, the newest earth stations of networks such as Intelsat join power stations, audio and TV broadcast stations and microwave relay systems in their ability to operate without interruption for long, if not indefinite, periods.

In the case of stations with low traffic requirements, particularly in remote areas or developing countries where experienced maintenance personnel are scarce, unattended reliable operation is of increasing importance. The Canadian domestic system was planned for unattended operation and has been operating since early 1973. It now has over 50 unattended stations.

Technological advances will affect several areas of commercial satellite earth stations. The antenna diameter will remain in the range of 30 m (100 feet) for the high-capacity Intelsat stations, with some stations using diameters as large as 32 m (105 feet) to permit the use of uncooled low-noise receivers with noise temperatures of about 50 K. The feed systems will require spectrum reuse capability with circular or rotatable linear polarization modes to work with the probable operational mode of the INTELSAT V satellite. Such a feed system needs an axial ratio for linear polarization to achieve the required spectrum reuse isolation in the range of 25 to 35 dB. Two such feed systems for the Telesat Canada heavy-route station at Allan Park, Ontario, and the Canadian West Coast station at Lake Cowichan, B.C., near Vancouver, are presently in use.

Time-division multiple-access (TDMA) and digital-signal modulation will be used in some of the next generation stations, particularly for the heavy route traffic in the Atlantic ocean community of nations. However, emphasis will be placed on the design evolution aspects which will make the station equipment more compact and easier to operate and maintain.

Major design changes will take place in the ground communications equipment (the GCE subsystem). This subsystem, consisting mainly of receiver and exciter chains, control consoles and the TV test and mounting facility, has been full solid-state since the inception of commercial earth stations. It contains a medium traffic-handling station of about ten racks of equipment. The next generation stations will be converted from discrete solid-state components and printed circuits in the baseband, IF and RF circuitry to integrated circuitry. This will bring about an approximate 10-fold reduction in the size of the modules and a 50% reduction in the number of racks. Such units as the modulator/upconverter and the downconverter will each consist of a flat alumina substrate measuring approximately 5.1 X 7.6 cm (2 X 3 in) upon which all components and wiring are printed. Reliability, production and test time will be significantly improved by this application of IC technology. Furthermore, IC's will reduce the size of the subsystem and the number of interconnecting cables, connectors and plugs. The mechanical configuration of the subsystem probably will be changed from the pullout shelves to modules or cards for front-panel insertion.

Another design improvement will be the replacement of the 40-dB TWT driver amplifier in the high-power-amplifier subsystem by a transferred-electron amplifier (TEA) with similar gain. This change, already instituted in some of the Phase 4 stations, will remove all electron tubes from earth stations except the ones used in the high-power amplifiers.

The use of small and large stations makes equalization of the radiated RF power necessary for efficient use of the satellite transponders. Since the network configuration is known and losses predictable, station transmitting power can be preassigned with calibration and adjustment cycles as needed. Other station functions such as antenna pointing and tracking can be performed automatically. Thus, new developments will include the expanded use of computers for monitoring and station control functions, such as switching and supervisory display of equipment status and displaying operational status of unattended stations.

References

1. James Warwick, "Providing Communications to Isolated Areas," *IEEE EASCON '73*, Conv. Record, Washington, D.C., October 1973.
2. P.M.M. Norman and D.E. Weese, "Thin Route Satellite Communications for Northern Canada," *IEEE International Conf. on Comm.*, Montreal, 1971.
3. W.K. Sones and L.E. Gray, "Design Features of an Unattended Earth Terminal for Satellite Communications," *IEEE ICC '73*, Conf. Record.
4. William F. Arnold, "Sales of Earth Terminals Set to Soar," A.D. Little, Inc.
5. Lloyd G. Ludwig, "Some Developments in Semi-Direct Broadcast Satellite and Community Receiving Systems," Proceedings of the 19th Annual Meeting of American Astronautical Society, AAS Paper No. 73-155, June 1973.
6. Andrew M. Werth and Dr. Tadahiro Sekimoto, "Small Earth Terminal Applications for Satellite Communications," Proceedings of the 19th Annual Meeting of American Astronautical Society, AAS Paper No. 73-121, June 1973.
7. J. Janky and J. Potter, "The ATS-F Health-Education Technology Communications System," *IEEE ICC 73*, Conf. Record.
8. A. Dickinson, "The Optimization of System Parameters for Minimum Cost Regional and National Satellite Systems," *IEEE Conference on Earth Station Technology*, London, October 1970.
9. B.G. Evans and R. Walters, "An Economic Satellite Communications System for Small Nations," *IEEE Int. Conf. on Comm.*, Montreal, 1971.
10. "Final Acts of the World Administrative Radio Conference for Space Telecommunications," published by the International Telecommunication Union, Geneva, 1971.
11. C.C. Han, K. Ohkubo, J. Albernaz, J.M. Janky, and B.B. Lusignan, "Optimization in the Design of a 12 GHz Low Cost Ground Receiving System for Broadcast Satellites," *IEEE International Conference on Communications,* Conf. Record, June 1973.
12. A.W. Brook, "An Inexpensive Earth Terminal for TV Reception in Bush Communities," *RCA Engineer*, vol. 19, No. 3, October/November 1973.
13. Bernard J. Mirowsky, "Design of a Cost Effective Satellite/Terrestrial Communication Network," *IEEE EASCON '73*, Washington, D.C., October 1973.
14. Lorne B. Dunn, "Telephony to Remote Communities in Canada, via Satellite (Using Single Channel per R.F. Carrier)," *IEEE International Conf. on Comm.*, Montreal, 1971.
15. H. Morikawa, M. Ogata, and A. Karakami, "A Newly Developed Intelsat Earth Station," *IEEE International Conf. on Comm.*, Conf. Record, Paper No. 8-30, June 1973.
16. J. Almond, "The Telesat Canada Domestic Communications Satellite System," Paper presented at CCITT Latin American Regional Plan Meeting, Brasilia, Brazil, July 1973.
17. Private Communications with Dr. Bruce Lusignan, Associate Professor of Electrical Engineering, the Graduate School of Stanford University.
18. Hygino Caetano Corsetti, "Telecommunications in Brazil," *Telecommunications Journal of Switzerland*, vol. 40, October 1973.
19. Charles C. Sanderson and Lloyd G. Ludwig, "Single Channel Per Carrier Voice Transmission via Communications Satellite," *AIAA 5th Communications Satellite Systems Conference*, Paper No. 74-71, Los Angeles, April 1974.
20. G.P. Petrick and C.M. Abrahamson, "Economic Considerations for Low-Capacity SHF Satellite Communications Earth Terminals" *AIAA 5th Communications Satellite Systems Conference*, Paper No. 74-459, Los Angeles, April 1974.

Part 5

Integration of Satellite
And Terrestrial
Communication Networks

Integration of Satellite And Terrestrial Communication Networks

Y. F. LUM
Bell-Northern Research
Ottawa, Canada

INTRODUCTION

Commercial communication via satellite was first introduced by Intelsat in 1965. Today the satellite communication network is worldwide and covers about 53 countries. In the USSR the Molniya domestic system has been in operation for some years. In the western world, the Canadian Telesat system, the first domestic satellite network, went into operation in 1973. This is now being followed by several domestic systems in the U.S.

Any satellite system, whether international or domestic, has to be integrated into and operate with existing communications networks (1). Today's networks provide the speed, ease and accuracy with which a person can be connected to any one of more than 300 million other telephone subscribers in almost any part of the world.

BACK HAUL INTERFACE

Because the current Intelsat and western world domestic satellites use frequency bands (4 GHz and 6 GHz) shared with terrestrial microwave systems, in most cases the earth stations have to be located at considerable distances from population centers. This necessitates back-haul links which may be provided by microwave systems using frequency bands other than the shared bands or by buried cable systems. In either case there are some interface problems.

Analog System Interface

Most current satellite systems use frequency modulation with an analog frequency-division-multiplexed baseband signal, referred to as FDM/FM. The modulated RF carrier is beamed to the satellite and, after the appropriate frequency translation and amplification by the transponder to a downlink signal, is transmitted to earth to be received by a number of earth stations. This is called a multi-destination carrier.

In the transmit direction, the back-haul and assembly of the FDM baseband channels, groups and supergroups are straightforward. (See Figure 1). The back-haul carries the FDM signal assembled at a toll center (or gateway international switching center) to the earth station, and interface at the earth station is at the FDM baseband.

In the receive direction, as shown in Figure 1, the situation is not as simple. First consider an earth station receiving a multi-destination carrier from a distant earth station. The carrier is demodulated and those supergroups, groups or channels preassigned to the particular receiving station are extracted by de-multiplexing equipment. Multi-destination carriers from other distant earth stations are processed in a similar manner. This results in a multiplicity of supergroups, groups and channels which must now be rearranged and recombined by multiplexing equipment to form a standard FDM baseband signal for transmission via the back-haul link to the toll center (or gateway). At the toll center, it could frequently happen that the transmit path of a two-way voice circuit would be located in a certain channel, group and supergroup and the receive path of the same circuit would be located in a different channel, group and supergroup, and the receive path of the nection would then be needed to pair the transmit and receive channels. This would incur a small risk of intelligible cross-talk.

The complexity of the back-haul multiplex arrangement can be reduced if the interface at the earth station is only at voice frequencies. To do this, both the satellite and the back-haul base-

131

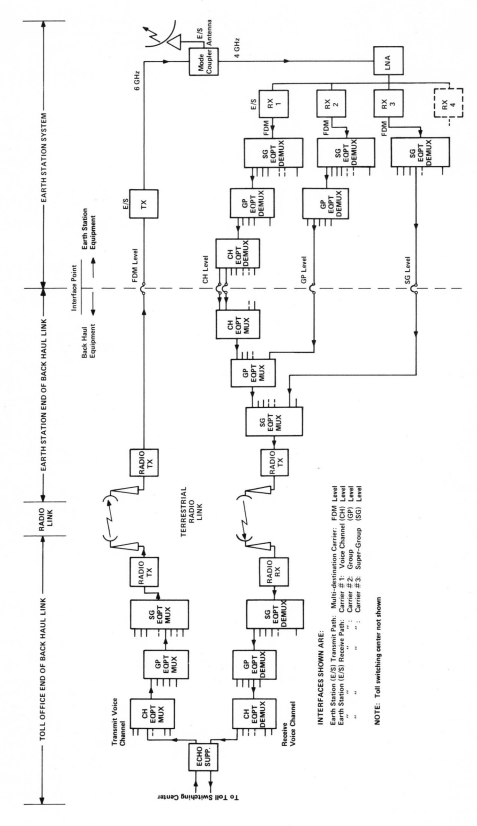

Figure 1. Analog (FDM/FM) Satellite System Interface

INTERFACES SHOWN ARE:

Earth Station (E/S) Transmit Path: Multi-destination Carrier: FDM Level
Earth Station (E/S) Receive Path: Carrier #1: Voice Channel (CH) Level
      "   "   " : Carrier #2: Group (GP) Level
      "   "   " : Carrier #3: Super-Group (SG) Level

NOTE: Toll switching center not shown

bands have to be translated down to voice frequency channels in both directions. The penalty is the added expense of the multiplexing and demultiplexing equipment. Another disadvantage is the degradation to circuit quality and reliability introduced by the additional channelling process.

With analog systems, one other multiplex interface problem is the supergroup and group pilot frequencies used either to monitor the system-level stability, or to regulate the level of the receive-end supergroup and group amplifiers. The pilot frequencies used in North America are different from those recommended by CCITT. Mainly because the North American pilot frequencies (104.08 kHz for group, 315.92 kHz for supergroup) are near the band edges, and therefore allow wideband data transmission, they have been adopted for most satellite systems.

Digital System Interface

The future satellite system may use a broadband digital technique, with time-division multiple-access (TDMA). The RF carriers containing the digital bursts would be multi-destinational. If the terminals of the satellite network use the same digital hierarchical structure (in North America, the T1 line, T2 line, etc.) then the interface at the earth stations could be directly digital at one or more of the digital hierarchical levels. Multiplexing problems similar to the analog case would still exist in the transmit and receive directions.

In a network where the satellite and terrestrial systems interface at digital levels, there is the further problem of synchronizing the two systems (2). Network synchronization is needed because the satellite uplink and downlink propagation path lengths are not constant but are constantly changing as the satellite drifts (the satellite does not remain perfectly stable at all times, although over a 24-hour period the drift error may be small). The satellite link, therefore, may appear at the interface points as an elastic store with changing storage capacity. Network synchronization discussed in the following is not to be confused with the problem of synchronizing the TDMA bursts of the uplink RF carriers. Network synchronization may be achieved in one of three ways:

1. By frame spilling

A whole frame is either suppressed or repeated depending on the relative speeds of the terrestrial and satellite system bit-timing clocks (3).

2. By elastic store

An elastic store at the receiving interface point could compensate for the changing capacity of the elastic storage of the satellite path.

3. By pulse stuffing

The satellite system digital bit-timing clock is allowed to run slightly faster than the terrestrial system clock. To make up for the difference, dummy pulses are inserted at the transmit earth station and removed at the receive earth station. Pulse stuffing in terrestrial PCM systems is an established technique.

With a telephone voice network any of the three techniques may be used. With data transmission the frame-spilling technique is not acceptable since vital data could be destroyed.

While digital transmission probably will have extensive future applications, there remains the complete lack of compatibility (except 8 kHz sampling) between the European and North American designs. Therefore, direct digital interconnection between the two systems, via satellite or any other means, is not currently feasible. Transcoding could provide a solution. Alternatively, the digital signals could be converted to voice channels and interfaced on an analog basis. The penalties would be cost, and quality and reliability of service.

SATELLITE TRANSMISSION DELAY

The characteristically long propagation-time delay of satellite systems causes integration problems. With geostationary satellites, the round-trip transmission-time delay is about 600 ms. This delay causes problems in echo suppression and signalling and some types of data transmission.

Echo Suppression

Because of the impossibility of achieving a perfect balance at all possible two-wire to four-wire interfaces, echoes are usually present in long-distance telephone circuits. If the transmission round-trip delay is more than about 45 ms, echo suppressors are required. Echo suppressors, therefore, are mandatory for satellite circuits (4). Current echo suppressor designs at best are compromises between the conflicting requirements of effective echo suppression on the one hand and permitting break-in or double-talk on the other hand. However, certain suppressors such as the canceller under development promise to be especially effective in satellite circuits. Echo suppression should be capable of being disabled by a tone when the circuit is used for data transmission.

Signalling

Signalling in international satellite networks is usually arranged by agreement between the administrations concerned. Currently most of the overseas connections are operator dialed. A widely used signalling system is the CCITT No. 5 and its variants which were developed for terrestrial systems. No major problems are encountered when this system is used on satellite connections. In the future the common-channel signalling CCITT No. 6 is likely to be used.

In the North American telephone network where subscribers usually dial their own calls, a number of minor signalling problems could arise with the switching machines currently in use. Trunk signalling equipment is either two-way or one-way. With two-way signalling, both ends of the circuit may independently initiate a circuit seizure giving rise to a blocked condition called "glare." Due to the long delay in transmitting a circuit-connect signal from one end to the other, the glare condition is expected to occur more frequently in satellite circuits than in landlines. The problem can be resolved either by using only one-way signalling or by equipping common-control switching machines with sender retrial facilities with the capability of recognizing a blocked condition, releasing the circuit and making another attempt. If only one-way signalling is employed, the penalty is a slight drop in traffic-handling efficiency. Other signalling problems are associated with the use of start-stop dialing or delay dialing in switching machines, but these are minor and can be easily resolved.

Tandem Satellite Circuits

Although two-hop satellite circuits have been used for certain remote area applications (such as the Thin-Route service in Canada) and during abnormal circumstances (such as restoration of submarine cable circuits by satellite after cable failure), in general such arrangements should be avoided for high-density toll-quality telephone service. In a domestic telephone network with its traditional hierarchical structure and alternative routing, some means should be provided to prevent a tandem satellite-circuit connection. The same applies for a domestic satellite call routed to an overseas destination, which may use satellite facilities. Special circuit arrangements could be used on a short-term basis to avoid a tandem connection. Long-term solutions may employ travelling class marks in the signalling system to identify the nature of the circuit and hence avoid a tandem satellite call.

Data Transmission

The long transmission delay has some effect on data transmissions that use a feedback system for error correction (5,6). In the low-speed stop-and-wait system in which a new block of data is not sent until an acknowledgement is received, there is a decrease in throughput efficiency when a satellite circuit is used. Apart from this, the data network can operate over satellite links. In some medium- and high-speed systems the data is sent continuously until a control signal in the return path indicates an error in a certain block, and the sender is requested to retransmit either the erroneous block or the erroneous block and all blocks thereafter. For such systems to operate over a satellite link, it is necessary to ensure that the sender's storage capacity is adequate to cover the long round-trip delay plus any transmission delay in the receive equipment. Otherwise, operation over the satellite system is not possible. For data transmission over a satellite link, forward error-correction schemes are best. Without a feedback system, the transmission delay of a satellite circuit is of no consequence.

DEMAND ASSIGNMENT SATELLITE NETWORK

Satellite networks may operate on a demand-assignment basis (7), instead of on preassigned circuits between points. Demand-assignment operation means that the circuit capacity is flexible from minute to minute and at all times is able to meet the traffic demands for which the network is designed. There are two basic demand-assignment configurations, depending on whether the satellite uses a single-channel-per-carrier frequency-division-multiple-access (SCPC/FMDA) system, such as the SPADE, or a multi-channel TDMA system.

In the SCPC/FDMA system, the network must interface on a channel basis at the earth stations. The network signalling must also be intercepted and buffered at the earth stations so that the circuit destination is known and a pair of satellite frequencies is assigned by a central processor unit to the originating and terminating earth stations to form a two-way circuit. At the end of the call, the frequency pair is returned to the common pool for use by other earth stations. This type of satellite demand-assignment system forms part of an overall switched network, with the switching function being performed within the satellite system. The key interface between satellite and terrestrial systems is therefore the network signalling which may either be individual channel signalling or common-channel inter-office signalling (CCIS).

In the multi-channel TDMA system, a different approach is used. The capacity of an earth station is given by the length of the uplink TDMA burst. As the length of some bursts increases, other bursts in the network must correspondingly decrease, so that the overall TDMA frame time remains constant. Channel capacity is thereby distributed to the earth stations on a bulk basis. The switching center maintains control of the individual circuit signalling and requests the earth station only to increase or decrease the bulk circuit capacities to the various distant earth stations. Consequently, interface is required only between a traffic-volume measuring device at the switching center and the capacity-allocation processor of the TDMA system at the earth station. In such a system, the satellite facility functions solely as a transmission medium.

SUMMARY

Integration between satellite and terrestrial communications networks concerns the following major interface areas:

1. The multiplex arrangements in the back-haul links
2. The satellite and terrestrial synchronization in digital networks
3. Satellite propagation time delays and the need for:
    a. Echo suppressors or echo cancellers
    b. Modifications to some common control switching equipment signalling
    c. Modifications to certain data transmission networks with feedback type error control
    d. Demand assignment operation of satellite system.

These areas represent problems whose solutions are available within the current state of the art.

References

1. S.C. Jenkins, "Integration of Satellite Circuits into the Telephone Network," U.S. Seminar on Comm. Sat. Earth Station Technology, 1966.
2. Y.F. Lum, "A Proposed Time Division Multiple Access (TDMA) Satellite System for Anik," *IEEE International Conf. on Comm.,* Seattle, 1973.
3. "Clock Synchronization between Incoming Digital (PCM) Systems and PCM/PSK/TDMA Satellite Transmission Systems," CCITT Com SP-D, Temp DOC 37-E, March 1971.
4. R.G. Gould and G.K. Helder, "Transmission Delay and Echo Suppression," *IEEE Spectrum,* vo. 7, no. 4, April 1970.
5. L.A. Cohen and G.V. Germano, "Gauging the Effects of Propagation Delay and Error Rate on Data Transmission Systems," *Telecommunications Journal,* vol. 37-VIII/1970.
6. M.D. Balkovic and P.E. Muench, "Effect of Propagation Delay, Caused by Satellite Circuits, on Data Communication Systems that use Block Retransmission for Error Correction," *IEEE International Conf. on Comm.,* June 1969.
7. J.G. Puente and A.M. Werth, "Demand Assignment Service for the Intelsat Global Network," *IEEE Spectrum*, vol. 8, no. 1, January 1971.

Addendum

Observation, Experimentation, Tracking and Navigation Satellite Systems

Observation, Experimentation, Tracking and Navigation Satellite Systems

JOSEPH DESKEVICH
Operations Research Inc.
Silver Spring, Maryland, U.S.A.

RALPH E. TAYLOR
National Aeronautics & Space Administration (NASA)
Goddard Space Flight Center
Greenbelt, Maryland, U.S.A.

During the 1960's, new techniques for earth observations, astronomy, and navigation have evolved as products of the space program conducted by the United States. By mid-1974, these techniques developed to the point where they served a major portion of the world's population with up-to-date satellite weather predictions, high-resolution imagery concerning the utilization of earth resources, and more accurate methods of accomplishing planetary and galactic scientific exploration. To discuss each of the satellites in detail is beyond the scope of this effort.

Consequently, this section is a compendium consisting of a brief description of each satellite program that is:

1. Currently being operated or planned for near future by NASA.

2. Launched by NASA and presently being operated by NOAA.

There is also a description of the Delta vehicle used to launch these satellites into orbit. Information for this section was obtained through the assistance of NASA, Goddard Space Flight Center, NOAA, and RCA, the prime contractor for development of the TIROS satellites.

The following tables are included as part of the program descriptions:

Table 1. Characteristics of NASA-Supported, Earth-Orbiting Satellites for Observation, Experimentation and Communication/Navigation

Table 2. Weather Satellite Programs

Table 2A. TIROS Mission Details

Table 2B. ESSA Mission Details

Table 2C. ITOS/NOAA Mission Details

Table 3. Weather Satellite Characteristics

Table 4. TIROS/ESSA/ITOS/NOAA Program Summary

Table 5. Delta Launch Vehicle Characteristics

Table 6. Delta Launch Vehicle History.

ATS-1, Applications Technology Satelitte (Geosynchronous)—Launched December 7, 1966. Apogee 35,839 km; perigee 35,737 km; period 1436 minutes; inclination angle ±5.6°. Spectrum utilization:

FREQUENCY (MHz)	EMISSION/POWER	PURPOSE
136.47 & 137.35	30F9/1.9 W	PCM/PM telemetry of spacecraft systems and experiment status on command, or PFM/PM telemetry of experimental data on command
VHF	30A9/3 kW	Earth to spacecraft telecommand
135.6	90F9/30 W	Spacecraft to earth VHF experiment link
149.22	90F9/5 kW	Earth to spacecraft VHF experiment link
149.22	90F/100 W	Aircraft and ship to space VHF experiment link
4119.599 & 4178.591	25000F2, F3, F4, F5, F9/2.5 W	Spacecraft to earth experimental communication data links
4135.946 & 4195.172	A0/2.5 W	Spacecraft tracking aids
6212.094 & 6301.050	25000F2, F3, F4, F5, F9/10 kW	Earth to spacecraft experimental communication data links

ATS-3, Applications Technology Satellite (Geosynchronous)—Launched November 5, 1967. Apogee 35,901 km; perigee 35,666 km; period 1435 minutes; inclination angle ±3.9°. Spectrum utilization:

FREQUENCY (MHz)	EMISSION/POWER	PURPOSE
136.47 & 137.35	30F9/1.9 W	PCM/PM telemetry of spacecraft systems and experiment status on command, or PFM/PM telemetry of experimental data on command
VHF	30A9/3 kW	Earth to spacecraft telecommand
135.6	90F9/30 W	Spacecraft to earth VHF experiment link
149.22	90F9/5 kW	Earth to spacecraft VHF experiment link
149.22	90F9/100 W	Aircraft and ships to spacecraft VHF experiment link
412.05	270F9/.1 W	Spacecraft to earth electron density measurement experiment; phase coherent with 137.35 MHz signal
4119.599	25000F2, F4, F5, F9/8 W	Spacecraft to earth experimental communication data link
4135.946	A0/2.5 W	Spacecraft tracking aid
6212.094	25000F2, F4, F5, F9/10 kW	Earth to spacecraft experimental communication data link

ATS-5, Applications Technology Satellite (Geosynchronous)—Launched August 12, 1969. Apogee 35,863 km; perigee 35,710 km; period 1436 minnutes; inclination angle ±1.3°. Spectrum utilization:

FREQUENCY (MHz)	EMISSION/POWER	PURPOSE
Same frequency complement as ATS-3, above, except no 135.6 and 149.22 MHz links but plus the following:		
1550.0	25000F2, F9/12 W	Spacecraft to earth experiment link
1565.82	A0/.3 W	Spacecraft experiment CW beacon on command
1651.02	25000F2, F9/1 kW	Earth to spacecraft experiment link
1550.48	2500F2, F4, F9/12 W	Spacecraft to earth experiment link
1651.5	2500F2, F4, F9/1 kW	Earth to spacecraft experiment link
1553.79	6000F2, F4, F5, F9/10 kW	Spacecraft to earth experiment link

Notes:

1. Emission/Power as defined in "Designation of Emissions," ITT "Reference Data for Radio Engineers," 5th Edition, pp. 1–16 to 1–18, 1970.
2. Bandwidth expressed in kilohertz except in megahertz when following letter "M".

ATS-5—Continued

FREQUENCY (MHz)	EMISSION/POWER	PURPOSE
6214.94	6000F2, F4, F5, F9/10 W	Earth to spacecraft experiment link
1650.55	100 F2, F4, F9/1 kW	Earth to spacecraft experiment link
4119.599	100 F2, F4, F9/8 W	Spacecraft to earth experiment link
4119.599	(Same as ATS-3, except add F3)	
6212.094	(Same as ATS-3, except add F3)	

ATS-6, Applications Technology Satellite (Geosynchronous)—Launched May 30, 1974. Apogee 35,844 km; perigee 35,798 km; period 1437 minutes; inclination angle ±1.8°. Spectrum utilization:

FREQUENCY (MHz)	EMISSION/POWER	PURPOSE
136.23 & 137.11	30F9/2 W	PCM/PM Telemetry on command
VHF	30A9/3 kW	Earth to spacecraft telecommand
20.008	AO/1 W	Spacecraft ITSA (NBS) ionospheric sounder beacon
39.0156	AO/.125 W	Spacecraft ITSA (NBS) ionospheric sounder beacon
39.91596	AO/.125 W	Spacecraft ITSA (NBS) ionospheric sounder beacon
40.016	AO/.5 W	Spacecraft ITSA (NBS) ionospheric sounder beacon
40.11604	AO/.125 W	Spacecraft ITSA (NBS) ionospheric sounder beacon
41.0164	AO/.125 W	Spacecraft ITSA (NBS) ionospheric sounder beacon
139.0556	AO/.065 W	Spacecraft ITSA (NBS) ionospheric sounder beacon
140.056	AO/.25 W	Spacecraft ITSA (NBS) ionospheric sounder beacon
141.0564	AO/.065 W	Spacecraft ITSA (NBS) ionospheric sounder beacon
359.144	AO/.022 W	Spacecraft ITSA (NBS) ionospheric sounder beacon
360.04396	AO/.04 W	Spacecraft ITSA (NBS) ionospheric sounder beacon
360.144	AO/.16 W	Spacecraft ITSA (NBS) ionospheric sounder beacon
360.24404	AO/.04 W	Spacecraft ITSA (NBS) ionospheric sounder beacon
361.144	AO/.022 W	Spacecraft ITSA (NBS) ionospheric sounder beacon
860	M40F3, F5/105 W	Spacecraft to earth (India only) TV link
1550	M12F2, F3, F9/40 W	Spacecraft to earth experiment link
1650	M12F2, F3, F9/1 kW	Earth to spacecraft experiment link
1650	M12F2, F3, F9/100 W	Aircraft to spacecraft experiment link
1659 MHz	M12F2, F3, F4, F9/40 W	Maritime ship to spacecraft experiment link
2075	M40F5, F9/20 W	Spacecraft to spacecraft data relay link (2062.85, ATS-6 to NIMBUS-F 1000F9/13 W) 2069. 1125, ATS-6 to GEOS-C, 400F9/13 W)
2075	M40F5, F9/20 W	Spacecraft to earth data relay link tests
2247	M20F5, F9/20 W	Spacecraft to spacecraft data relay link (GEOS-C to ATS-6, 400F9/5 W)
2250	M40F5, F9/50 W	Earth to spacecraft data relay link tests
2253	M20F5, F9/20 W	Spacecraft to spacecraft data relay link (NIMBUS-F to ATS-6, 6000F9/2, 4 or 8 W)
2569.2 & 2670	M40F5/15 W	Spacecraft to earth experimental TV links
3950	M500F9/20 W	Spacecraft to earth RFI experimental link
3750, 3950 & 4150	M40F2, F3, F4, F5, F9/12 W	Spacecraft to earth experimental communication data links
5950, 6150 & 6350	M40F2, F3, F4, F5, F9/10 kW	Earth to spacecraft experimental communication data links
5925—6425	AO/5 W	Earth to spacecraft RFI experimental links
6301.05	M25F9/100 W	Earth (mobile) to spacecraft RFI experimental link

ATS-6 (F)—Continued

FREQUENCY (MHz)	EMISSION/POWER	PURPOSE
13,190–13,200 & 17,790–17,800	AO/25 W	Earth to spacecraft millimeter wave propagation experimental links
19,000–21,000 & 29,000–31,000	AO/10 W	Spacecraft to earth millimeter wave propagation experimental links

NIMBUS-4, Meteorological Observation Satellite—Launched April 8, 1970. Apogee 1098 km; perigee 1088 km; period 107 minutes; inclination angle 99.8°. Spectrum utilization:

FREQUENCY (MHz)	EMISSION/POWER	PURPOSE
136.5	30A9/.5 W	Continuous radiation of PCM/AM telemetry of realtime data or PCM/AM telemetry of stored data on command
136.95	30F9/5 W	AM/FM telemetry of APT data as programmed
VHF	30A9/3 kW	Earth to spacecraft telecommand
401.5	300F9/25 W	IRLS meteorological data on command
466.0	100F9/25 W	Earth to spacecraft IRLS data
1702.5	3000F9/2 W	Meteorological sensor data on command

NIMBUS-5, Meteorological Observation Satellite—Launched December 11, 1972. Apogee 1102 km; perigee 1088 km; period 107 minutes; inclination angle 99.9°. Spectrum utilization:

FREQUENCY (MHz)	EMISSION/POWER	PURPOSE
136.5	30A9/.5 W	Continuous radiation of PCM/AM telemetry of realtime data or PCM/AM telemetry of stored data on command
VHF	30A9/3 kW	Earth to spacecraft telecommand
1702.5	3000F9/4 W	Meteorological sensor data on command
2208.5	3000F9/4 W	Surface composition mapping radiometer data on command

NIMBUS-F, Meteorological Observation Satellite—Launched October, 1974. Apogee 1100 km; perigee 1100 km; period 107 minutes; inclination angle 100°. Spectrum utilization:

FREQUENCY (MHz)	EMISSION/POWER	PURPOSE
136.5	30A9/.5 W	Continuous radiation of PCM/AM telemetry of realtime data or PCM/PM telemetry of stored data on command
VHF	30A9/3 kW	Earth to spacecraft telecommand
401.2	30F9/1 W	Balloon to spacecraft data link (TWERLE)
1702.5	3000F9/4 W	Meteorological sensor data on command
2062.85	1000F9/10 kW	Earth to spacecraft ranging data link
2062.85	1000F9/13 W	ATS-6 to NIMBUS ranging data link
2253	6000F9/2, 4 or 8 W	Spacecraft to ATS-F and/or earth stations ranging data link

ERTS-1, Earth Resources Technology Satellites—Launched July 23, 1972. Apogee 914 km; perigee 901 km; period 103 minutes; inclination angle 99°. Spectrum utilization:

FREQUENCY (MHz)	EMISSION/POWER	PURPOSE
137.86	3F9/90F9/.5 W	Continuous radiation of narrow-band PCM/PM telemetry; wide-band PCM/PM telemetry on command

ERTS-1—Continued

FREQUENCY (MHz)	EMISSION/POWER	PURPOSE
VHF	30A9/3 kW	Earth to spacecraft telecommand
401.55	100F9/12.6 W	Earth to spacecraft PSK/PM data links (data collection platforms)
2106.4	3600F9/10 kW	Earth to spacecraft USB data link
2229.5	20000F9/20 W	Spacecraft to earth multidetector radiometric scanner data link
2265.5	20000F9/20 W	Spacecraft to earth multispectral TV camera data link
2287.5	5000F9/1 W	Spacecraft to earth USB data link

ERTS-B, Earth Resources Technology Satellite—To be launched first quarter 1975. Apogee 918 km; perigee 918 km; period 104 minutes; inclination angle 99°. Spectrum utilization same as ERTS-1.

ISIS-1, International Satellite for Ionospheric Studies—Launched January 30, 1969. Apogee 3513 km; perigee 575 km; period 128 minutes; inclination angle 88°. Spectrum utilization:

FREQUENCY (MHz)	EMISSION/POWER	PURPOSE
0.150 to 20.0	PO/4 W	Sweep-frequency topside ionospheric sounder signals; PW 100 μs/PRF 30/400 W pk pwr or PW 100 μs/PRF 30 or 60/100 W pk pwr as programmed by telecommand. Rate of frequency sweep is approximately 1 MHz.
0.250, 0.480, 1.0, 1.950, 4.0 & 9.303	PO/4 W	Fixed-frequency topside sounding signals; PW, PRF and pk pwr options same as above. Transmission period is 3–5 seconds.
136.08	100F9/4 W	PAM/FM telemetry of sounder data on command
136.41	AO/.1 W	Continuous radiation of CW beacon
136.59	50F9/3 W	PCM/PM telemetry of auxiliary data on command
137.95	AO/.1 W	CW beacon for DRTE tracking
VHF	30A9/3 kW	Earth to spacecraft telecommand
401.75	300F9/4 W	AM/PCM/PM/FM telemetry on command
401.75	500F9/4 W	AM/PCM/PM/FM telemetry on command for acquisition only

ISIS-2, International Satellite for Ionospheric Studies—Launched April 1, 1971. Apogee 1422 km; perigee 1356 km; period 113 minutes; inclination angle 88°.
Spectrum utilization same as ISIS-1, except 0.120 MHz in lieu of 0.250 MHz for fixed-frequency topside sounding signals.

SMS-1,B,C, Synchronous Meteorological Satellites—SMS-1 launched May 17, 1974. SMS-B to be launched first quarter 1975. SMS-C to be launched first quarter 1976. Apogee 35,700 km; perigee 35,700 km; period 1440 minutes; inclination angle ±1°. Spectrum utilization:
NOTE: GEOS-A in same orbit with SMS-C.

FREQUENCY (MHz)	EMISSION/POWER	PURPOSE
135.565 & 137.195	40F9/2 W	Range and range rate data on command
136.38	90F9/2 or 8 W	PCM/PM and/or PAM/FM/PM telemetry with high or low power on command.
VHF	30A9/3 kW	Earth to spacecraft telecommand
401.85	300F9/10 W	Data collection platform links to spacecraft (NOAA)
402.082081 & 402.088081	1F9/40 W	Balloon to spacecraft data links (GARP)
468.825	150F9/10 or 40 W	Spacecraft link to data collection platforms

SMS-1, B, C—Continued

FREQUENCY (MHz)	EMISSION/POWER	PURPOSE
1682.5	25000F9/20 W	Spacecraft to earth data link; functions include ranging, WEFAX, DCP reports PCM/FM/PM telemetry and/or VISSR information.
2028.0	6000F9/100 W	Data collection platform links to spacecraft (NOAA)
2030.5	11000F9/1 kW	Earth to spacecraft data link; functions include ranging, WEFAX, DCP interrogation, telecommand and/or VISSR information.

OAO-C, Orbiting Astronomical Observatory (COPERNICUS)—Launched August 21, 1972. Apogee 749 km; perigee 739 km; period 100 minutes; inclination angle 35°. Spectrum utilization:

FREQUENCY (MHz)	EMISSION/POWER	PURPOSE
136.26	30F9/2 W	PCM/PM telemetry on command
136.44	AO/.16 W	Continuous radiation of CW beacon
VHF	30A9/3 kW	Earth to spacecraft telecommand
400.55	300F9/7 W	FM/FM, PCM/FM (slow digital) or PCM/FM (fast digital) telemetry on command

OSO-7, Orbiting Solar Observatory—Launched September 29, 1971. Apogee 392 km; perigee 289 km; period 91 minutes; inclination angle 33°. Spectrum utilization:

FREQUENCY (MHz)	EMISSION/POWER	PURPOSE
136.29	30F9/90F9/.57 W	Continuous radiation of narrowband realtime PCM/PM telemetry; wideband stored data PCM/PM telemetry on command
VHF	30A9/3 kW	Earth to spacecraft telecommand

OSO-I, Orbiting Solar Observatory—Launched March 1975. Apogee 550 km; perigee 550 km; period 96 minutes; inclination angle 33°. Spectrum utilization:

FREQUENCY (MHz)	EMISSION/POWER	PURPOSE
136.92	30F9/2 W	Continuous radiation of PCM/PM telemetry
VHF	30A9/3 kW	Earth to spacecraft telecommand
2212.5	1000F9/5 W	PCM/PM telemetry on command

IUE, International Ultraviolet Explorer (Geosynchronous)—To be launched fourth quarter 1976. Apogee 36,300 km; perigee 36,300 km; period 1440 minutes; inclination angle 28°. Spectrum utilization:

FREQUENCY (MHz)	EMISSION/POWER	PURPOSE
136.86	90F9/6 W	PCM/PM telemetry on command
136.145 & 137.575	40F9/1.2 W	Range and range rate data on command
VHF	40A9/3 kW	Earth to spacecraft telecommand
2071.691	3600F9/1 kW	Earth to space craft USB data link
2249.8	3600F9/2 W	Spacecraft to earth USB data link

SAS-1, Small Astronomy Satellite (EXPLORER 42)—Launched December 12, 1970. Apogee 536 km; perigee 504 km; period 95 minutes; inclination angle 3°. Spectrum utilization:

FREQUENCY (MHz)	EMISSION/POWER	PURPOSE
136.68	3F9/90F9/.25 or 1.5 W	Continuous radiation of narrow-band PCM/PM telemetry for tracking and "housekeeping" data acquisition; wide-band PCM/PM telemetry of stored data on command
VHF	30A9/3 kW	Earth to spacecraft telecommand

SAS-2, Small Astronomy Satellite (EXPLORER 48)—Launched November 15, 1972. Apogee 615 km; perigee 440 km; period 95 minutes; inclination angle 1.9°. Spectrum utilization same as SAS-1.

SAS-C, Small Astronomy Satellite—Launched May 1975. Apogee 550 km; perigee 550 km; period 96 minutes; inclination angle 0°. Spectrum utilization:

FREQUENCY (MHz)	EMISSION/POWER	PURPOSE
136.68	3F9/90F9/.25 or 1.5 W	Continuous radiation of narrowband PCM/PM telemetry for tracking and "housekeeping" data acquisition; wideband PCM/PM telemetry of stored data on command
VHF	30A9/3 kW	Earth to spacecraft telecommand
2250.0	100F9/.5 or 4 W	PCM/PM telemetry of experimental data on command for earth station acquisition or ATS-F data relay

IMP-6, Interplanetary Explorer (EXPLORER 43)—Launched March 13, 1971. Apogee 197,100 km; perigee 8126 km; period 5968 minutes; inclination angle 34°. Spectrum utilization:

FREQUENCY (MHz)	EMISSION/POWER	PURPOSE
136.17	30F9/4 W	FM/PM telemetry of special purpose analog data on command
135.428 & 136.912	40F9/.5 W	Range and range rate data on command
137.17	30F9/8 W	Continuous radiation of PCM/PM telemetry for tracking and realtime data acquisition
VHF	30A9/3 kW	Earth to spacecraft telecommand

IMP-7, Interplanetary Explorer (EXPLORER 47)—Launched September 23, 1972. Apogee 249,689 km; perigee 181,856 km; period 17,367 minutes; inclination angle 7.4°. Spectrum utilization:

FREQUENCY (MHz)	EMISSION/POWER	PURPOSE
136.89	30F9/8 W	FM/PM telemetry of experiment and tracking data or backup PCM/PM telemetry on command
136.208 & 137.572	40F9/.9 W	Range and range rate data on command
137.92	30F9/12 W	Continuous radiation of PCM/PM telemetry
VHF	30A9/3 W	Earth to spacecraft telecommand

IMP-8, Interplanetary Explorer (EXPLORER 50)—Launched October 26, 1973. Apogee 289,303 km; perigee 145,380 km; period 17,551 minutes; inclination angle 28°. Spectrum utilization:

FREQUENCY (MHz)	EMISSION/POWER	PURPOSE
136.8	30F9/8 W	FM/PM telemetry of experiment and tracking data or backup PCM/PM telemetry on command

IMP-8—Continued

FREQUENCY (MHz)	EMISSION/POWER	PURPOSE
136.02 & 137.58	40F9/.9 W	Range and range rate data on command
137.98	30F9/12 W	Continuous radiation of PCM/PM telemetry
VHF	30A9/3 kW	Earth to spacecraft telecommand

HAWKEYE, Neutral Point Explorer (Ionospheric)—Launched June 3, 1974. Apogee 95,600 km; perigee 200 km; period 1920 minutes; inclination angle 90°. Spectrum utilization:

FREQUENCY (MHz)	EMISSION/POWER	PURPOSE
136.29	30F9/1 W	Continuous radiation of PCM/PM telemetry
VHF	30A9/3 kW	Earth to spacecraft telecommand
400.65	300F9/1 W	PCM/PM telemetry on command

BE-C, Beacon Explorer (EXPLORER 27)—Launched April 29, 1965. Apogee 1322 km; perigee 933 km; period 108 minutes; inclination angle 41°. Spectrum utilization:

FREQUENCY (MHz)	EMISSION/POWER	PURPOSE
136.74	AO/30F9/.4 W	Continuous radiation of CW tracking beacon or PAM/FM/PM telemetry
VHF	30A9/3 kW	Earth to spacecraft telecommand

AE-C, Atmosphere Explorer (EXPLORER 51)—Launched December 16, 1973. Apogee 4295 km; perigee 150 km; period 132 minutes; inclination angle 68°. Spectrum utilization:

FREQUENCY (MHz)	EMISSION/POWER	PURPOSE
137.23	AO/.25 W/90F9/1 W	CW beacon or PCM/PM telemetry on command
2108.25	3000F9/10 kW	Earth to spacecraft USB data link
2289.5	3000F9/5 W	Spacecraft to earth USB data link

AE-D, Atmosphere Explorer—Launched March, 1975. Apogee 4000 km; perigee 150 km; period 129 minutes; inclination angle 100°. Spectrum utilization same as AE-C.

OV5-6, Orbiting Vehicle—Launched May 23, 1969. Apogee 122,463 km; perigee 6046 km; period 3114 minutes; inclination angle 27°. Spectrum utilization:

FREQUENCY (MHz)	EMISSION/POWER	PURPOSE
136.38	3F9/.5 W	Continuous radiation of PAM/FM/PM telemetry
400.45	AO/.75 W	CW beacon as programmed

CAS-A, Cooperative Applications Satellite (EOLE)—Launched August 16, 1971. Apogee 903 km; perigee 676 km; period 101 minutes; inclination angle 50°. Spectrum inclination:

FREQUENCY (MHz)	EMISSION/POWER	PURPOSE
136.35	30F9/.25 W	Continuous radiation of PCM/PM realtime telemetry or stored data on command
VHF	30A9/3 kW	Earth to spacecraft telecommand
401.71796	10F9/1 or 3 W	EOLE experiment balloon to spacecraft link
464.486	22F9/4 W	EOLE experiment spacecraft to balloon link

SM-C2, San Marco Satellite (International)—Launched February 18, 1974. Apogee 833 km; perigee 243 km; period 96 minutes; inclination angle 3°. Spectrum utilization:

FREQUENCY (MHz)	EMISSION/POWER	PURPOSE
136.74	AO/30F9/.5 W	CW beacon or PAM/FM/PM telemetry on command
VHF	30A9/3 kW	Earth to spacecraft telecommand

INTASAT, Spanish Satellite (International)—Launched July 1974. Apogee 1460/1520 km; perigee 1460/1520 km; period 115 minutes; inclination angle 102°. Spectrum utilization:

FREQUENCY (MHz)	EMISSION/POWER	PURPOSE
40.01 & 41.01025	AO/.2 W	Ionospheric sounder beacons
136.71	15F9/.1 W	Continuous radiation of PCM/PSK/PM telemetry

UK-5, International Satellite (Astronomy)—Launched August 1974. Apogee 480 km; perigee 480 km; period 94 minutes; inclination angle 3°. Spectrum utilization:

FREQUENCY (MHz)	EMISSION/POWER	PURPOSE
137.68	30F9/.08 W	Continuous radiation of PCM/PM telemetry of realtime data stored on command
VHF	30A9/3 kW	Earth to spacecraft telecommand

HEOS-A2, Highly Elliptical Orbiting Satellite (ESRO)—Launched January 31, 1972. Apogee 236,255 km; perigee 4499 km; period 7485 minutes; inclination angle 88°. Spectrum utilization:

FREQUENCY (MHz)	EMISSION/POWER	PURPOSE
136.68	30F9/6 W	Continuous radiation of PCM/PSK/PM telemetry and/or ranging data on command
VHF	30A9/3 kW	Earth to spacecraft telecommand

AEROS-B, German Aeronomy Satellite (International)—Launched July 1974. Apogee 1000 km; perigee 230 km; period 97 minutes; inclination angle 97°. Spectrum utilization:

FREQUENCY (MHz)	EMISSION/POWER	PURPOSE
137.29	AO/30F9/.15 W or 90F9/1.5 W	Continuous radiation of CW beacon or PCM/PM telemetry; wideband PCM/PM telemetry of stored data on command
VHF	30A9/3 kW	Earth to spacecraft telecommand

MTS, Meteoroid Technology Satellite (EXPLORER 46)—Launched August 13, 1972. Apogee 787 km; perigee 494 km; period 98 minutes; inclination angle 38°. Spectrum utilization:

FREQUENCY (MHz)	EMISSION/POWER	PURPOSE
136.32	30F9/.5 W	PCM/PM telemetry on command
136.65	10F9/.075 W	Continuous radiation of PCM/PM telemetry, but only when spacecraft is in sunlight
VHF	30A9/3 kW	Earth to spacecraft telecommand

GEOS-2, Geodetic Research Satellite (EXPLORER 36)—Launched January 11, 1968. Apogee 1574 km; perigee 1080 km; period 112 minutes; inclination angle 106°. Spectrum utilization:

FREQUENCY (MHz)	EMISSION/POWER	PURPOSE
136.32	20F9/.4 W	Continuous radiation of PAM/FM/PM telemetry with special function modes on command
VHF	30A9/3 kW	Earth to spacecraft telecommand
162,324 & 972	AO/.5 W	Spacecraft doppler transmitter
224.25 & 449	1000F9/1.4 W	Spacecraft to earth geodetic data links
420.9	1000F9/1 kW	Earth to spacecraft geodetic data links
5690	PO/1 MW	Earth to spacecraft tracking radar
5765	PO/400 W	Spacecraft radar transponder

GEOS-C, Geodetic Research Satellite—Launched October 1974. Apogee 843 km; perigee 843 km; period 102 minutes; inclination angle 115°. Spectrum utilization:

FREQUENCY (MHz)	EMISSION/POWER	PURPOSE
136.32	30F9/.4 W/90F9/1.5 W	Continuous radiation of narrowband PCM/PM telemetry with wideband PCM/PM telemetry on command
VHF	30A9/3 kW	Earth to spacecraft telecommand
162 & 324	5F2/.4 W	Spacecraft doppler beacons (TRANET)
2069.1125	1000F9/2 kW	Earth to spacecraft data link
2069.1125	400F9/20 W	Spacecraft (ATS-F) to spacecraft data link
2247	400F9/5 W	Spacecraft to spacecraft (ATS-F) data relay; spacecraft to earth data link
5690	PO/1 MW	Earth to spacecraft tracking radars
5765	PO/400 W	Spacecraft radar transponder
13,900	M120P9/2 kW	Spacecraft radar altimeter
27,800	M240P9/160 W	Spacecraft radar altimeter (2nd harmonic)
41,700	M360P9/40 W	Spacecraft radar altimeter (3rd harmonic)

SYNCOM-3, (Geosynchronous)—Launched August 19, 1964. Apogee 35,831 km; perigee 35,799 km; period 1438 minutes; inclination angle ±8°. Spectrum utilization:

FREQUENCY (MHz)	EMISSION/POWER	PURPOSE
136.98	40F9/2 W	PAM/FM/PM telemetry on command
VHF	30A9/3 kW	Earth to spacecraft telecommand

TOS-F, ESSA-8, Operational TIROS Meteorological Observation Satellite—Launched December 15, 1968. Apogee 1462 km; perigee 1412 km; period 115 minutes; inclination angle 102°. Spectrum utilization:

FREQUENCY (MHz)	EMISSION/POWER	PURPOSE
136.77	30F9/.25 W	Continuous radiation of FM/PM telemetry for tracking and "housekeeping" data acquisition
137.62	30F9/5 W	AM/FM telemetry of APT data; continuous radiation as programmed
VHF	22A9/1 kW	Earth to spacecraft telecommand

ITOS-D, NOAA-2, Improved TIROS Meteorological Observation Satellite—Launched October 15, 1972.
Apogee 1454 km; perigee 1448 km; period 115 minutes; inclination angle 102°. Spectrum utilization:

FREQUENCY (MHz)	EMISSION/POWER	PURPOSE
136.77	30F9/.25 W	Continuous radiation of FM/PM telemetry for tracking and "housekeeping" data acquisition; realtime SPM and/or VTPR engineering data on command
137.5 & 137.62	30F9/5 W	AM/FM telemetry of realtime SR data; continuous radiation on either channel selectable by command
VHF	22A9/1 kW	Earth to spacecraft telecommand
1697.5	3000F9/5 W	FM/FM telemetry of realtime VHRR data; stored VHRR, VTPR, SR, SPM or "housekeeping" data on command

ITOS-F, NOAA-3, Improved TIROS Meteorological Observation Satellite—Launched November 6, 1973.
Apogee 1509 km; perigee 1500 km; period 116 minutes; inclination angle 102°.
Spectrum utilization same as ITOS-D except 137.14 MHz in place of 136.77 MHz; continuous radiation on either channel selectable by command.

ITOS-G, Improved TIROS Meteorological Observation Satellite—Launched July 1974; Apogee 1460 km;
perigee 1460 km; period 115 minutes; inclination angle 102°. Spectrum utilization same as ITOS-F.

SSS-A, Small Scientific Satellite (EXPLORER 45)—Launched November 15, 1971. Apogee 25,077 km;
perigee 267 km; period 436 minutes; inclination angle 3.4°. Spectrum utilization:

FREQUENCY (MHz)	EMISSION/POWER	PURPOSE
136.83	3F9/.5 W	Continuous radiation of PCM/PM telemetry
137.95	AO/3F9/90F9/3 W	Wideband FM/PM telemetry or backup narrowband PCM/PM telemetry of CW beacon on command
VHF	30A9/3 kW	Earth to spacecraft telecommand

ANS-A, Astronomical Netherlands Satellite—Launched August 1974. Apogee 550 km; perigee 450 km;
period 95 minutes; inclination angle 98°. Spectrum utilization:

FREQUENCY (MHz)	EMISSION/POWER	PURPOSE
137.89	AO/10F9/.25 W or 30F9/2 W	Continuous radiation of CW beacon or narrowband PCM/PM telemetry; wideband PCM/PM telemetry on command
VHF	30A9/3 kW	Earth to spacecraft telecommand

INTELSAT II-F1, Synchronous Communications Satellite (Commercial)—Launched October 26, 1966.
Apogee 37,153 km; perigee 3235 km; period 718 minutes; inclination angle 18°. Spectrum
utilization:

FREQUENCY (MHz)	EMISSION/POWER	PURPOSE
136.44 & 136.98	AO/40F9/2 W	CW beacon or PAM/FM/FM telemetry on command
4058–4184	30000F9/36 W	Spacecraft to earth communications data links
4058.15 & 4182.0	30F9/1 W	Spacecraft tracking aids
6283–6409	30000F9/12.5 kW	Earth to spacecraft communications data link

INTELSAT IV-F6 & F8, Synchronous Communications Satellites (Commercial)—F8 launched June 1974.
F6 launched September 1974. Apogee 35,788 km; perigee 35,779 km; period 1440 minutes; inclination angle 0.2°. Spectrum utilization:

FREQUENCY (MHz)	EMISSION/POWER	PURPOSE
3700–4200	36000F9/20 W	Spacecraft to earth communications data links (12 transponders with 36 MHz maximum bandwidth; 4 MHz guard band between transponders)
5925–6425	36000F9/12.5 kW	Earth to spacecraft communications data links

TABLE 2. Weather Satellite Programs

TIROS (Television and IR Observation Satellite)

TIROS was the first generation of weather satellites and demonstrated the feasibility of collection and transmission of meteorological data by satellite. They photographed cloud cover by TV camera and sensed atmospheric temperatures by IR radiometry.

Ten satellites of the series were launched. The early versions were spin-stabilized; torquing capability was added in later versions. TIROS 9 (forerunner of ESSA) was a cartwheel configuration with spin axis normal to the orbital plane. The series produced more than 500,000 cloud cover photographs and provided early hurricane watch.

Goddard Space Flight Center (GSFC) provided project and launch vehicle management. Radio Corporation of America (RCA) was the prime spacecraft contractor.

Table 2A shows mission details of the TIROS series satellites.

TABLE 2A. TIROS Mission Details

Satellite Designation and Launch Date	Orbit 1. Apogee (km) 2. Perigee (km) 3. Inclination 4. Period (min)	Launch 1. Vehicle 2. Weight (kg) 3. Range	Mission Resumé
TIROS-1 (1960 B2) Apr. 1, 1960	693 755 48.3° 99.2	Thor-Able 119 Eastern	World's first weather satellite. Sent 22,952 photos to June 1960. Useful life 89 days. Carried 2 1.27 cm Vidicons—one 1450 × 1450 km coverage and one 890 × 890 km coverage.
TIROS-2 (1960-1) Nov. 23, 1960	625 730 48.5° 98.3	Delta 126 Eastern	Sent 36,156 photos to December 1961. Useful life 376 days. Vidicons same as TIROS-1. Carried 5-channel wideband scanning IR radiometer; magnetic torquing.
TIROS-3 (1961 P1) July 12, 1961	736 824 47.9° 100.3	Delta 129 Eastern	First hurricane observation satellite. Useful life 230 days. Vidicons same as TIROS-1. Sent 35,033 photos to February 1962. Carried 3 1R radiometers, Univ. of Wisconsin radiometer experiment.
TIROS-4 (1962 B1) Feb. 8, 1962	704 849 48.3° 100.3	Delta 130 Eastern	First international data use. Useful life 161 days. Sent 32,593 photos to June 1962. Vidicons same as TIROS-1. Three IR radiometers: 1 scanning, 2 non-scanning.
TIROS-5 (1962 AA1) June 19, 1962	583 979 58.1° 100.5	Delta 130 Eastern	Hurricane watch. Useful life 321 days. Sent 58,226 photos to May 1963. Vidicons and radiometers same as TIROS-1. Inclination increased 10° to extend observation to higher latitudes.

Satellite Designation and Launch Date	Orbit 1. Apogee (km) 2. Perigee (km) 3. Inclination 4. Period (min)	Launch 1. Vehicle 2. Weight (kg) 3. Range	Mission Resumé
TIROS-6 (1962 AX1) Sept. 18, 1962	681 716 58.3° 98.7	Delta 127 Eastern	Hurricane watch, Mercury MA-8 and MA-9 support. Sent 68,557 photos to October 1963. Useful life 389 days. Vidicons and radiometers same as TIROS-1.
TIROS-7 (1963 24A) June 19, 1963	617 654 58.2° 97.4	Delta 136 Eastern	Hurricane watch. Useful life 1809 days; deactivated February 1966. Sent 125,331 photos. Two 1.27 cm Vidicons with 1200 × 1200 km coverage, ion temperature probe, one 5-channel scanning radiometer, two wide-angle low-resolution resistance thermometers.
TIROS-8 (1963 54A) Dec. 21, 1963	704 752 58.5° 99.4	Delta 118 Eastern	First use of automatic picture transmission (APT) for direct readout TV. Sent 102,463 photos to July 1967. Useful life 1287 days. Vidicon same as TIROS-7; APT 2.54 cm, 1320 × 1320 km coverage.
TIROS-9 (1965 4A) Jan. 22, 1965	707 2582 96.4° 119.2	Delta 136 Eastern	Produced first world cloud-cover photomosaic; daily global coverage; 16 pictures per orbit at 128 sec. intervals. Useful life 1238 days. Sent 88,892 photos. Deactivated Feb. 1967. Had two 1.27 cm Vidicons canted 26.5° on opposite sides of plane of rotation, 500-line, wide-angle coverage; 2 tape recorders, 48 pictures each. One camera failed, April 1965.
TIROS-10 (1965 51A) July 2, 1965	745 837 98.6° 100.7	Delta 132 Eastern	Hurricane watch. Sent 78,824 photos. Useful life 730 days. Two Vidicons same as TIROS-1 mounted in base. Spin axis controlled to 20° N latitude for hurricane watch.

TOS/ESSA (TIROS Operational System/Environmental Science Services Administration)

The ESSA-series satellites were developed from TIROS technology. Stored pictures in the advanced Vidicon camera subsystem (AVCS), provided global meteorological coverage on a daily basis. Automatic picture transmission (APT) permitted realtime readout.

The ESSA provided program management; GSFC provided spacecraft and launch vehicle management; and RCA was prime spacecraft contractor.

Table 2B shows mission details of the ESSA series satellites.

TABLE 2B. ESSA Mission Details

Satellite Designation and Launch Date	Orbit 1. Apogee (km) 2. Perigee (km) 3. Inclination 4. Period (min)	Launch 1. Vehicle 2. Weight (kg) 3. Range	Mission Resumé
ESSA-1 (1966 8A) Feb. 3, 1966	707 840 97.9° 100.3	Delta 138 Eastern	Part of world's first meteorological system. Sent 111,144 photos. Useful life 861 days. Turned off May 1967.

Satellite Designation and Launch Date	Orbit 1. Apogee (km) 2. Perigee (km) 3. Inclination 4. Period (min)	Launch 1. Vehicle 2. Weight (kg) 3. Range	Mission Resumé
ESSA-2 (1966 16A) Feb. 28, 1966	1358 1414 101.0° 113.5	Tad 135 Eastern	With ESSA-1 completed first meteorological system. Sent 126,631 photos. Useful life 1692 days. First operational realtime APT, consisting of two 2.54 cm Vidicons, 800 line, 3150 × 3150 km coverage with 3.7 km resolution. Transmit pictures every 352 sec.
ESSA-3 (1966 87A) Oct. 2, 1966	1387 1490 101.1° 114.6	Tad 149 Western	Replaced ESSA-1. Sent 92,076 photos. Life 738 days. Vidicons same as ESSA-2. Dual tape recorders 48 frames, 6 or 12 photos per orbit, stored for U.S. readout. Turned off October 1968.
ESSA-4 (1967 6A) Jan. 26, 1967	1328 1424 102.0° 113.4	Tad 135 Western	Replaced ESSA-2. Sent 27,169 photos. Vidicons same as ESSA-2. One camera failed third day; remaining camera malfunction caused deactivation December 1967. Life 314 days.
ESSA-5 (1967 36A) Apr. 20, 1967	1358 1424 101.9° 113.5	Tad 149 Western	Replaced ESSA-3. Sent 86,715 photos. Useful life 1034 days. Vidicons and AVCS same as ESSA-2. Radiation sensors measured global heat balance.
ESSA-6 (1967 114A) Nov. 10, 1967	1410 1488 102.1° 114.8	Tad 135 Western	Replaced ESSA-4. Sent 64,164 photos. Life 763 days. Vidicons same as ESSA-2. Deactivated November 1969.
ESSA-7 (1968 69A) Aug. 16, 1968	1432 1475 101.7° 114.9	Long-tank Delta 149 Western	Replaced ESSA-5. Sent 57,000 photos. Life 570 days. Vidicons and radiation sensors same as ESSA-5. S-band transmission. Recorders failed. Deactivated July 1969.
ESSA-8 (1968 114A) Dec. 15, 1968	1416 1465 101.8° 114.6	Long-tank Delta 135 Western	Realtime APT to worldwide receivers. Sent 148,300 photos. Life 1502 days. Still operational January 1973.
ESSA-9 (1969 16A) Feb. 26, 1969	1427 1507 101.7° 115.2	Taid 149 Eastern	Replaced ESSA-7. Sent 156,500 photos. Life 1322 days. S-band transmission. Still operational January 1973.

ITOS/NOAA (Improved TIROS Operational System/National Oceanic and Atmospheric Administration)

A second generation system for global weather observation was provided in ITOS/NOAA. A single spacecraft contained the combined capability of two ESSA satellites. Vertical temperature profiles obtained by flat-plate radiometry provided first three-dimensional weather information. Measurements included land and sea surface and cloud-type temperatures. The spacecraft were earth-pointing, attitude-stabilized with deployable solar panels. ITOS E, F and G will be launched before the end of 1975.

NOAA provided project management; GSFC provided spacecraft and launch vehicle management. RCA was the prime spacecraft contractor.

Table 2C shows mission details of the ITOS/NOAA satellites.

Satellite Designation and Launch Date	Orbit 1. Apogee (km) 2. Perigee (km) 3. Inclination 4. Period (min)	Launch 1. Vehicle 2. Weight (kg) 3. Range	Mission Resumé
ITOS-1 (TIROS-M) 1970 8A) Jan. 23, 1970	1435 1482 102.0° 115.0	Tat Delta M 307 Western	Second generation operational meteorological satellite-operational June 1970. Sent 227,000 photos. Life 510 days. Two AVCS 2.54 cm Vidicons, 800-line, 3150 × 3150 km coverage, 3.7 km resolution; photos at 260 sec. intervals; 2 tape recorders; 3-orbit storage.
NOAA-1 (ITOS-A) (1970 106A) Dec. 11, 1970	1429 1482 101.9° 114.8	TTAT Delta N 307 Western	Meteorological satellite. Sent 72,000 photos. Life 252 days. Two APT 2.54 cm Vidicons, 600-line, 3150 × 2400 km coverage, 3.7 km resolution; pictures at 260 sec. intervals.
ITOS-B (1971 91A) Oct. 21, 1971	293 1483 102.5° 102.7	Delta N 310 Western	Failed to achieve correct orbit due to launch vehicle malfunction. Two scanning radiometers, visible 0.5 to 0.73 µ, 3.7 km resolution, 1R 10.51 to 12.5 µ, 7.4 km resolution.
NOAA-2 (ITOS-C) (1972 82A) Oct. 15, 1972	1451 1458 101.7° 114.9	Delta N 338 Western	First operational temperature recording sensor. First very-light-resolution radiometer (0.9 km). Johns Hopkins APL solar proton monitor measures proton/electron flux−60, 30, 10 Mev protons, 100 to 750 Kev electrons, 12 to 32 Kev alpha particles. Univ. of Wisconsin flat-plate radiometer 0.3 to 30 µ.

TABLE 3. Characteristics of Weather Satellites

TIROS
   DESCRIPTION (Dimensions in cm)
      18-sided cylinder:
         Diameter−106.7
         Height−55.9
      Solar cells around periphery

   WEIGHT BREAKDOWN (kg)

TIROS	1	8	9
Structure[1]	41.0	41.0	41.8
Electronics (comm, T/M, & command)	11.4	11.4	12.7
Electric power	20.5	20.9	20.5
Attitude control	1.3	4.1	7.7
Harness and RF cabling	9.1	9.1	11.4
Payload (sensors, recorders, TV transmitters)	36.4	31.8	42.3
TOTAL[3]	119.7	118.3	136.4

   CONTROL SYSTEM
      TIROS 1−Spin stabilized at 10 rpm. Ten 0.45-kg spin-up rockets for spin control.
      TIROS 2-8−Spin stabilized similar to TIROS plus magnetic torquing for limited 3-axis control.
      TIROS 9−Spin stabilized axis normal to orbit plane. Magnetic torquing for spin control. Liquid passive damper control system to maintain attitude ±1° in all 3 axes.

TABLE 3. Characteristics of Weather Satellites — Continued

ELECTRICAL POWER
    Solar Array
- 9120 silicon solar cells, 1 × 2 cm. P/N Type TIROS 1; N/P type TIROS 2-10
- Total array area 1.65 sq. meters
- Power Output—55 W max.; 30 W orbit average

    Battery
- 63.4-Ah Ni-Cad cells.

TELEMETRY
    Subcarriers
- TIROS 1—One 1.3 Hz
- TIROS 8—One 1.3 Hz
- TIROS 9[2]—Two 1.3 and 2.3 Hz

    Beacon Transmitter
- TIROS 1—One 108 MHz
- TIROS 8—Two 136.23 and 136.92 Hz
- TIROS 9[2]—Two 136.23 and 136.92 Hz

    Time-Division Multiplexing
- TIROS 1—One 40-point S/W
- TIROS 8—Two 40-point S/W
- TIROS 9[2]—Two 90-point solid-state T/M commutator

COMMAND SYSTEM
    TIROS 1 & 8
- 10 audio tones
- 148 MHz (nominal)

    TIROS 9 & ESSA 1
- Binary-coded FSK commands as well as 10 audio tones
- 148 MHz (nominal)
- Command verification

TOS/ESSA
    DESCRIPTION (Dimensions in cm)
        18-sided cylinder:
            Diameter—106.7
            Heights—55.9
        Solar cells around periphery

WEIGHT BREAKDOWN (kg)

ESSA[2]	2, 4, 6, 8	3, 5, 7, 9
Structure[1]	41.8	44.6
Electronics (Comm, T/M, and command)	16.4	17.3
Electronic power	24.1	18.2
Attitude control	8.7	8.2
Harness and RF cabling	6.8	13.6
Payload (sensors, recorders, TV transmitters)	31.8	46.5
TOTAL[3]	129.6	148.4

CONTROL SYSTEM
    ESSA 1—Same as TIROS 9
    ESSA 2-9—Spin-stabilized axis normal to orbit plane. Roll and yaw axis magnetically controlled. Magnetic torquing for spin control. Liquid passive damper control system to maintain ±1° in all 3 axes.

ELECTRICAL POWER
    Solar Array (ESSA 2-9)[2]
- 10,020 silicon solar cells, 1 × 2 cm, N/P type
- Total array area 2.02 sq. meters
- Power Output—55 W max.; 40 W orbit average

    Batteries
- (ESSA 2, 4, 6, 8)[2] Three 4-amp. hr., each with 21 Ni-Cad cells
- (ESSA 3, 5, 7, 9) Two 4-Ah, each with 21 Ni-Cad cells

TABLE 3. Characteristics of Weather Satellites — Continued

TOS/ESSA — Continued

TELEMETRY

Subcarriers

- ESSA 2, 4, 6, 8–Three 2.3, 3.0, 3.9 kHz on each transmitter
- ESSA 3, 5, 7, 9–Same as ESSA 2 but T/M also recorded on video tape during picture taking

Beacon Transmitter

- ESSA 2, 4, 6, 8–Two 136.77 MHz
- ESSA 3, 5, 7, 9–Same as ESSA 2; S-Band 1.7 GHz added to ESSA 7 and 9

Time-Division Multiplexing

- ESSA 2-9–Two 90-point solid state T/M commutators for housekeeping. Spin rate and attitude continuously telemetered.

COMMAND SYSTEM

ESSA 2-9[2]

- Redundant command chain
- Binary-coded FSK commands
- 4 audio tones used to activate command subsystem
- 148 MHz (nominal)

ITOS/NOAA

DESCRIPTION (Dimensions in cm)

Box shaped:

Base–101.6 × 101.6

Height–121.9

Solar cells on three extendable panels.

Solar panel deployed span–428.

WEIGHT BREAKDOWN (kg)

NOAA	1	2
Structure	57.2	65.5
Electronics (comm, T/M, and command)	21.8	19.1
Electric power[1]	51.3	54.0
Attitude control	30.9	34.6
Harness and RF cabling	20.9	25.0
Thermal control and misc.	20.5	13.6
Payload (sensors, recorders, processors, controllers, TV transmitters)		
TOTAL[3]	309.6	340.3

CONTROL SYSTEM

- 3-axis stabilized within ±1° in all 3 axes.
- Main body despun to 1 rev. per orbit. Momentum wheel on pitch axis spins at 150 rpm.
- Roll and yaw axes magnetically controlled.
- Magnetic control for momentum adjustment.
- Passive liquid dampers for nutation damping.

ELECTRICAL POWER

Solar Array (All Versions)

- 10,260 N/P type solar cells, 2 × 2 cm
- Total array area of 3 deployable panels–4.1 sq. meters
- Power Output–400 W max.; 290 W orbit average

Battery

- (1 TOS 1, NOAA 1)
- Two 4-amp. hr., each with 23 Ni-Cad cells
- (NOAA 2)
  Two 6-Ah, each with 23 Ni-Cad cells

TELEMETRY

ITOS 1 and NOAA 1

- Analog–Two subcarriers 2.3 and 3.9 kHz on two beacon transmitters 136.77 MHz
- S-Band communications 1.7 GHz
- PAM time-division multiplexing; two 120-point solid-state T/M commutators
- Selected T/M data digitized and recorded for later transmittal

NOAA 2–Same as ITOS 1 except:
- 200-point commutators
- All T/M data digitized and recorded as well as analog real-time trnsmission

COMMAND SYSTEM
    All ITOS and NOAA
- 132 commands, twice ESSA capacity
- Binary-coded FSK commands
- 148 MHz (nominal)

(1) Including solar array   (2) ESSA 1 same as TIROS 9   (3) Not including adapter weight

TABLE 4. TIROS/ESSA/ITOS/NOAA Program Summary

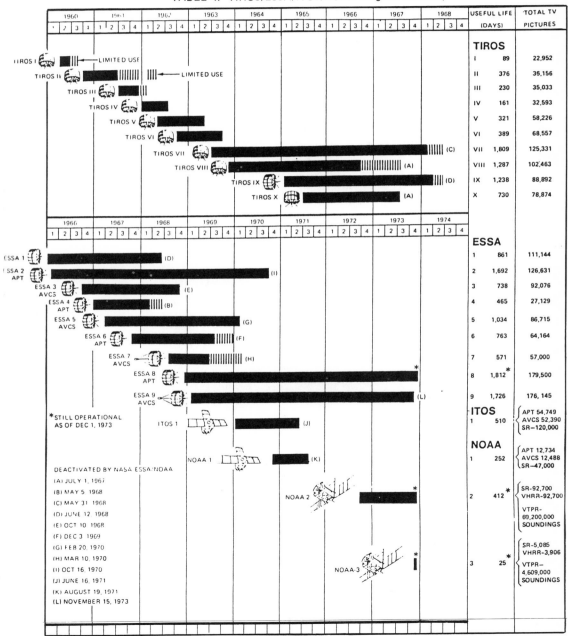

TABLE 5. Delta Launch Vehicle Characteristics

General Characteristics:

Height:	35.4 meters (116 feet) (includes shroud)
Maximum diameter:	2.4 meters (8 feet) (without attached solids)
Lift-off weight:	106,000 kilograms (about 116 tons)
Lift-off thrust:	1.4 million newtons (305,000 pounds) (including strap-on solids)

First Stage: (Liquid only) extended long tank Thor

Diameter:	2.4 meters (8 feet)
Height:	27 meters (90 feet) including interstage
Propellants:	RP-1 kerosene is used as the fuel and liquid oxygen (LOX) is utilized as the oxidizer.
Thrust:	900,000 newtons (205,000 pounds)
Burning time:	About 3 minutes and 50 seconds
Weight:	Approximately 84,700 kilograms (93 tons)
Strap-on solids:	Three solid propellant rocket motors.
Diameter:	0.8 meters (31 inches each)
Height:	0.6 meters (1.96 feet each)
Weight:	4,500 kilograms (4.9 tons each)
	13,410 kilograms (29,568 pounds)
Thrust:	321,317 newtons (52,000 pounds each)
Burning time:	38 seconds

Second Stage:

Propellants:	Liquid-Aerozene 50 for the fuel and nitrogen tetroxide (N_2O_4) for the oxidizer.
Diameter:	1.5 meters (5 feet) plus 2.4 meters (8 feet) attach ring.
Height:	6.1 meters (20 feet)
Weight:	6210 kilograms (6.8 tons)
Thrust:	42,000 newtons (9,400 pounds)
Burning time:	314 seconds-first burn; 18 seconds-second burn

Third Stage:

Propellants:	Solids
Height:	1.5 meters (5 feet)
Diameter:	1 meter (3 feet)
Weight:	723 kilograms (1593 pounds)
Thrust:	42,300 newtons (9,500 pounds)
Burning time:	44 seconds

Payload Launch Capability

Delta 2914:

1500 pounds, synchronous transfer orbit.
4400 pounds, near-earth orbit, 100 nm, circular.

Delta 2910 Castor IV (2-Stage):

1940 pounds, synchronous transfer orbit
5200 pounds, near-earth orbit.

TABLE 6. Delta Launch Vehicle History

Delta No.	Payload	Weight (lbs.)	Launch Date	Launch Time	Pad	Vehicle
1	ECHO	180	051360	0440 EDT	17A	DM-19
2	ECHO 1A	182	081260	0439 EDT	17A	DM-19
3	TIROS A2	277	112360	0613 EST	17A	DM-19
4	EXPL-X (P-13)	79	032561	1017 EST	17A	DM-19
5	TIROS A3	285	071261	0525 EDT	17A	DM-19
6	EXPL-XII (S-3)	84	081561	2221 EDT	17A	DM-19
7	TIROS 4 (D)	285	020862	0743 EST	17A	DM-19
8	PSP 1 (S-16)	458	030762	1106 EST	17A	DM-19
9	ARIEL (S-51UK1)	136	042662	1800 GMT	17A	DM-19
10	TIROS 5 (F)	286	061962	0719 EDT	17A	DM-19
11	TELSTAR 1 (TSXI)	171	071062	0335 EDT	17B	DM-19
12	TIROS 6 (F)	280	091862	0353 EDT	17A	DM-19
13	EXPL XIV (S-3A)	86	100262	1711 EDT	17B	DSV-3A
14	EXPL XV (S-3B)	98	102762	1815 EST	17B	DSV-3A
15	RELAY A-15	170	121362	1830 EST	17A	DSV-3B
16	SYNCOM A-25	146	021463	0035 EST	17B	DSV-3B
17	EXPL XVII (S-6)	410	040263	2100 EST	17A	DSV-3B
18	TELSTAR 2 (TSX2)	176	050763	0638 EDT	17B	DSV-3B
19	TIROS 7 (G)	296	961963	0450 EDT	17B	DSV-3B
20	SYNCOM B (A-26)	147	072663	0933 EDT	17A	DSV-3B
21	EXPL XVIII (IMP A)	138	112663	2130 EST	17B	DSV-3C
22	TIROS 8 (H)	265	122163	0430 EST	17B	DSV-3B
23	RELAY II (A-16)	184	012164	1614 EST	17B	DSV-3B
24	S-66	132	031964	0613 EST	17A	DSV-3B
25	SYNCOM C	145	081964	0715 EDT	17A	DSV-3D
26	IMP B	135	100364	2245 EDT	17A	DSV-3C
27	S 3C	101	122164	0400 EST	17A	DSV-3C
28	TIROS I (EYE)	301	012265	0252 EST	17A	DSV-3C
29	OSO B2	547	020365	1136 EST	17B	DSV-3C
30	COMSAT HS303A	149	040665	1847 EST	17A	DSV-3D
31	IMP C	128	052965	0700 EDT	17B	DSV-3C
32	TIROS OT 1	280	070165	1106 EDT	17B	DSV-3C
33	OSO C	625	082565	1017 EDT	17B	DSV-3C
34	GEOS A	387	110665	1338 EDT	17A	DSV-3E
35	PIONEER A	146	121665	0231 EST	17A	DSV-3E
36	OT 3	304	020366	0241 EST	17A	DSV-3C
37	OT 2	286	022866	0858 EST	17B	DSV-3E
38	AE B	492	052566	0900 EDT	17B	DSV-3C1
39	IMP D	212	070166	1102 EDT	17A	DSV-3E1
40	PIONEER B	138	081766	1020 EDT	17A	DSV-3E1
41	TOS A	316	100266	0339 PDT	SLC2E	DSV-3E
42	INTELSAT II A (F-1)	355	102666	1805 EST	17B	DSV-3E1
43	BIOS A	950	121466	1420 EST	17A	DSV-3G
44	INTELSAT II B (F-2)	357	011167	0555 EST	17B	DSV-3E1
45	TOS B	285	012667	0932 PST	SLC-2E	DSV-3E
46	OSO E1	600	030867	1112 EST	17A	DSV-3C
47	INTELSAT II C (F-3)	365	032267	2030 EST	17B	DSV-3E1
48	TOS C	327	042067	0421 PDT	SLC2E	DSV-3E
49	IMP F	163	052467	0705 PDT	SLC2E	DSV-3E1
50	AIMP E	230	071967	1019 EDT	17B	DSV-3E1
51	BIOS-B	955	090767	1804 EDT	17B	DSV-3G
52	INTELSAT II D (F-4)	357	092767	2045 EDT	17B	DSV-3E1
53	OSO-D	605	101867	1158 EDT	17B	DSV-3C1
54	TOS-D	299	111067	0953 PST	SLC 2E	DSV-3E1
55	PIONEER C	146	121367	0908 EST	17B	DSV-3E1
56	GEOS B	469	011168	0816 PST	SLC 2E	DSV-3E1
57	RAE-A	602	070468	1026 PDT	SLC 2E	DSV-3J
58	TOS-E	347	081668	0524 PDT	SLC 2E	Delta N
59	INTELSAT III A (F-1)	641	091868	1809 EDT	17A	Delta M
60	PIONEER D	147	110868	0446 EST	17B	DSV-3E1
61	HEOS A	237	120568	1855 GMT	17B	DSV-3E1
62	TOS-APT F	297	121568	0921 PST	SLC 2E	Delta N
63	INTELSAT III C (F-2)	642	121868	1932 EST	17A	Delta M
64	OSO-F	645	012269	1148 EST	17B	DSV-3C1

158

TABLE 6. Delta Launch Vehicle History—Continued

Delta No.	Payload	Weight (lbs.)	Launch		Pad	Vehicle
			Date	Time		
65	ISIS A	532	013069	0646 GMT	SLC 2E	DSV-3E1
66	INTELSAT III B (F-3)	642	020569	1939 EST	17A	Delta M
67	TOS G	347	022669	0247 EST	17B	DSV-3E1
68	INTELSAT III D (F-4)	647	052169	2200 EDT	17A	Delta M
69	IMP G	175	062169	0147 PDT	SLC 2W	DSV-3E1
70	BIOS D	1546	062869	2316 EDT	17A	Delta N
71	INTELSAT III E (F-5)	647	072569	2206 EDT	17A	Delta M
72	OSO-G	647	080969	0352 EDT	17A	Delta N
73	PIONEER-E	148	082769	1759 EDT	17A	Delta L
74	SKYNET A	535	112269	0037 GMT	17A	Delta M
75	INTELSAT III F (F-6)	647	011470	1916 EST	17A	Delta M
76	TIROS M	682	012370	0331 PST	SLC 2W	Delta N6
77	NATO A	535	032070	0652 EST	17A	Delta M
78	INTELSAT III G (F-7)	647	042270	1946 EST	17A	Delta M
79	INTELSAT III H (F-8)	647	072370	1923 EDT	17A	Delta M
80	SKYNET B	535	081970	1211 GMT	17A	Delta M
81	ITOS-A	680	121170	1135 GMT	SLC 2W	Delta N6
82	NATO-B	533	020271	2041 EST	17A	Delta 3L-1
83	IMP-I	635	031371	1615 GMT	17A	Delta 3L-1
84	ISIS-B	570	040171	0257 GMT	SLC 2E	Delta 3E1
85	OSO-H	1416	092971	0945 GMT	17A	Delta N
86	ITOS-B	687	192171	1132 GMT	SLC 2E	Delta N6
87	HEOS A2	260	013172	1720 GMT	SLC 2E	Delta L
88	TD 1A	1043	031272	0155 GMT	SLC 2E	Delta N
89	ERTS A	270	072372	1806 GMT	SLC 2W	D 0900
90	IMP H	860	092372	0120 GMT	17B	D 1604
91	ITOS-D	742	101572	1719 GMT	SLC 2W	D 0300
92	TELSAT A	1238	111072	0114 GMT	17B	D 1914
93	NIMBUS F	1574	121172	0756 GMT	SLC 2W	D 0900
94	TELESAT B	1238	042073	2347 GMT	17B	D 1914
95	RAE-B	734	061073	1413 GMT	17B	D 1913
96	ITOS-E	747	071673	1710 GMT	SLC 2W	D 0300
97	IMP-J	876	102673	0226 GMT	GMT 17B	D 1604
98	ITOS-F	746	110673	1702 GMT	SLC 2W	D 0300
99	AE-C	1494	121673	0618 GMT	SLC 2M	D 1900

Index

American Satellite Corp., 50
    earth stations, 50
        communications antenna, characteristics of, 50
Anik, 33, 70
Antenna, 39–41, 45, 50, 52, 93
    AT&T earth stations, 41
    American Satellite Corp. earth station, 50
    COMSAT, 39–41
    General Electric Co. earth stations, 52
    polarization, orthogonal, 93
    Western Union earth stations, 45
Antenna radiation pattern, 103
Antenna system, GTE, 42
Applications Technology Satellite (*see* ATS-6)
AT&T earth stations, 41
    antennas, 41
    communications capacity, 41
    frequencies, 41
    transmitters, 41
ATS-6, 3–12

Back haul interface, 131–133
    analog system, 131–133
    digital system, 133
Broadcast Satellite Service, 73–77

Canada (*see* TELESAT Canada; *see also* CTS)
Canadian Domestic System (*see* TELESAT Canada)
Channel performance, TELESAT, 70–71
Communication networks (*see* Satellite, communication
        networks)
Communication satellite (*see* Satellite)
Communications, maritime (*see* Satellite)
Communications, optical, 90–91
Communications capacity, AT&T earth station, 41
Communications services, TELESAT Canada, 34
Communications Technology Satellite (*see* CTS)
Communications transmitter, GTE, 42
COMSAT, 38–41
    satellites, 38–41
        antenna, 39–41
        EIRP, 41
        flux density, 41
        launch vehicle, 41
        MM-wave propagation experiment, 41
        station-keeping functions, 41

COMSAT General Corp. (*see* COMSAT)
Coordination of geostationary satellite systems (*see*
        Satellite systems)
CTS, 13–18
    mission design features, 13–18
        communications experiments, 14–15
        communications subsystem, 14
        spacecraft, description, 14
    user program, Canada, 15–18
        and US, 18

Data transmission, 71, 134
    TELESAT, 71
Delta Launch Vehicle, 157–159
    characteristics, 157
    history, 158–159
Demand assignment satellite network, 134
Domestic communication satellite systems (*see* Satellite)

Echo suppression, 133
EIRP, 41, 113
    COMSAT, 41
European Regional Satellite Organization (*see* EUROSAT)
EUROSAT, 55–56
ESSA, 151–152, 156
    mission details, 142–143, 151–152
    program summary, 156
Earth station, 11–12, 20, 27, 34, 41–45, 49–52, 70,
        94–95, 103, 113, 121–128
    American Satellite Corp., 50
    antenna radiation pattern, 103
    AT&T, 41
    COMSAT, 41
    EIRP, 113
    General Electric Co., 50–52
    GTE Satellite Corp., 42
    HET, Denver, 11–12
    INTELSAT, 27
    orthogonal polarization (frequency reuse), 94–95
    RCA Global/Alaska Communications, Inc., 49–50
    SYMPHONIE project, 20
    technology, 121–128
        design and development, 124–128
            evolution of for operation with INTELSAT
                network, 124
            review of, 124–128
        systems planning, 121–124

161

US domestic communication satellite systems (*see* Satellite)

USSR (*see* MOLNIYA/ORBITA)

USSR domestic system (*see* MOLNIYA/ORBITA)

Voice performance, TELESAT, 70

Weather satellites, 150–156

WARC-ST 1971, 113–117
   coordination distance, changes in method for calculating 114–117
   conditions of use, changes in, 113
   earth station EIRP, 113
   power flux-density limits, 113

WARC-ST 1971—Continued
   results of, 113
   table of frequency allocations, changes to, 113

West Germany (*see* SYMPHONIE project)

Western Union Telegraph Co., 42–45
   earth stations, 42–45
      antenna, 45
      maximum modulating frequency, 43
      polarization, 42–43
      transmitters, 43–45
   satellites, 42

World Administrative Radio Conference for Space Telecommunications 1971 (*see* WARC-ST 1971)